ANIMAL MECHANICS

BIOLOGY SERIES
General Editor: R. Phillips Dales
Reader in Zoology in the University of London at Bedford College

The Biology of Estuarine Animals
J. Green

Structure and Habit in Vertebrate Evolution
G. S. Carter

Animal Ecology
Charles Elton, F.R.S.

Animal Evolution
G. S. Carter

The Biology of the Sea-shore
F. S. Flattely and C. L. Walton

A General Zoology of the Invertebrates
G. S. Carter

A Hundred Years of Evolution
G. S. Carter

The Nature of Animal Colours
H. Munro Fox, F.R.S., and H. Gwynne Vevers

Vertebrate Zoology
G. R. de Beer, F.R.S.

The Feathers and Plumage of Birds (Monograph)
A. A. Voitkevich

IN PREPARATION:

Developmental Genetics and Animal Patterns
K. C. Sondhi

Description and Classification of Vegetation
David W. Shimwell

Marine Biology
H. Friedrich

Principles of Histochemistry
W. G. Bruce Casselman

The Investigation of Natural Pigments
G. Y. Kennedy

Practical Invertebrate Zoology
F. E. G. Cox, R. P. Dales, J. Green, J. E. Morton,
D. Nichols, D. Wakelin

A photograph by Dr C. J. Pennycuick of a fulmar (*Fulmarus*) gliding. This photograph is discussed on page 246. (From Pennycuick and Webbe, 1959)

ANIMAL MECHANICS

by

R. McNeill Alexander

Lecturer in Zoology
at the
University College of North Wales
Bangor

UNIVERSITY OF WASHINGTON PRESS
SEATTLE

Printed in Great Britain

ACKNOWLEDGEMENTS

I AM grateful to Dr J. D. Currey for sending me typescripts of several papers while they were still unpublished, and to Dr C. J. Pennycuick for supplying the frontispiece.

About half of the illustrations have been reproduced from papers and other books. The sources are indicated in the captions, by references to the bibliography. I am grateful for permission to reproduce illustrations to the many authors and to:

The American Physiological Society, the Anthropological Society of Nippon, the Cambridge Philosophical Society, the Company of Biologists, the Linnaean Society of London, the Council of the Marine Biological Association of the United Kingdom, the Royal Society, the Zoological Society of London, and the Wistar Institute.

The Editorial Committee of the *Journal of Anatomy*, the Editorial Board of the *Journal of Physiology* and the Editors of *Acta Oto-laryngologia*, the *British Medical Bulletin*, *British Birds*, the *New Scientist*, *School Science Review*, and *Science*.

The Academic Press; Baillière, Tindall and Cassell Ltd; the Cambridge University Press; the Clarendon Press; W. H. Freeman & Co; Longmans, Green & Co; the McGraw-Hill Book Co; Springer-Verlag, and John Wiley & Sons Inc.

CONTENTS

PREFACE

THIS book is about the parts of mechanics that have been useful in zoology, and about some of the zoological investigations in which they have been used. I have not tried to mention every zoological paper that involves mechanics but have chosen topics for their interest, and to illustrate the widest possible range of mechanical principles.

I suppose that the average reader will be a zoologist who abandoned physics before he left school, and who knows no engineering. I have therefore assumed very little prior knowledge of mechanics and have tried to make my explanations simple. I have not thought it necessary to give mathematical proofs of the equations included in the book, but have stated where proofs can be found.

Each chapter deals with a major branch of mechanics. Each consists of alternating sections on mechanics and zoology: a section on a mechanical topic is followed by one or more sections on zoological topics involving the mechanical one, before the next mechanical topic is introduced. This means that related zoological topics involving different branches of mechanics may be widely separated. In such cases, I have put cross-references in the text.

CHAPTER 1

FORCE AND ENERGY

THIS chapter is about statics and dynamics, and their applications in zoology. A lot of the basic physics is elementary and easy, and most readers will wish merely to be reminded of it, rather than have it explained to them. Explanations can be found in school textbooks such as Nelkon and Parker (1965).

Forces and motion

The concept of force is basic to this chapter and indeed to the whole of this book. The definition of force is implicit in Newton's three Laws of Motion.

1. A body remains at rest or moves at a constant velocity unless forces act on it.

2. An unbalanced force gives a body an acceleration in the direction of the force. This acceleration is proportional to the force and inversely proportional to the mass of the body.

3. If body A exerts a force on body B, body B exerts an equal and opposite force on body A.

The second law is used to define the unit of force. A dyne is the force needed to give a mass of 1 gram an acceleration of 1 centimetre per second per second (i.e. to increase its velocity by 1 cm/s every second). A force of F dyn gives a mass of m gm an acceleration a cm/s² where

$$F = ma \tag{1}$$

Objects falling unimpeded under the influence of gravity have an acceleration of 981 cm/s² so that the force due to gravity on a body of mass m is 981 m dyn. The acceleration due to gravity is customarily represented by the letter g, so the force can be written mg dyn. It can also be given as m gram weight.

There is another way of expressing the second law. The momentum of a body is the product of its mass and its velocity, so *ma* is the rate at which its momentum is changing. The force (in dynes) acting on a body is equal to its rate of change of momentum (in g cm/s²).

There are various methods of measuring forces. Sometimes, a force can be measured simply by balancing it against a weight or by finding how much it will stretch a spring. Sometimes forces can be calculated from other observed quantities such as accelerations, as we shall see when we discuss jumping animals. Sometimes it is best to use one of the many electrical devices known collectively as force transducers (Donaldson, 1958). These respond to forces by producing electrical signals which can be amplified if necessary, and displayed on a cathode ray oscilloscope, a pen recorder or an ultra-violet recorder. When forces which change rapidly are to be measured, the design of the apparatus has to be considered very carefully, as we shall see in Chapter 7 (page 283).

To describe a force, one must give its direction as well as its size. In other words, force is a vector quantity like velocity. This must be remembered when forces are added together. An example will serve as a reminder of the method of adding vectors.

Locusts jump by rapidly extending their long hind legs. There is a good deal about locust legs and jumping in subsequent sections of this book: we will consider the forces which act in the leg during take-off (page 25), the relationship between these forces and the strength of the leg (page 159) and the arrangement of joints and muscles in the leg (page 37). It will be convenient to solve now a problem in vector addition which will arise later.

Only the hind legs of a locust are in contact with the ground as it takes off at the beginning of a jump. They exert forces on the ground because they are supporting the locust's weight and because they are accelerating it. If the mass of the locust is *m* g, the force due to its weight is 981 *m* dyn, and it of course acts vertically downwards. We will see on page 26 that the acceleration requires a force of about 14,500 *m* dyn, at an angle of about 55° to the horizontal.

Fig. 1a shows how these two forces can be added to obtain their resultant. AB is a vertical line drawn as nearly as possible 981 units long. BC is a line 14,500 units long, inclined at 55° to the horizontal. These lines represent the two forces and the

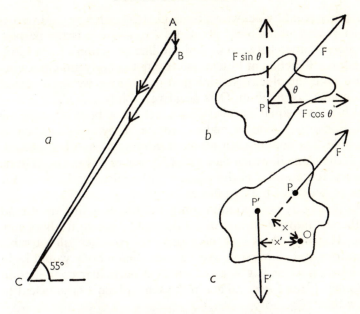

Figure 1. This figure is explained in the text

arrows show the directions of the forces. Notice that AB and BC have been arranged so that the arrows can be followed through from A to C. The length and direction of AC give the resultant force. It proves to be 15,300 m dyn, acting at 57° to the horizontal.

Not only can forces be added to obtain resultants: they can also be resolved into components acting in various directions. For instance, Fig. 1*b* shows a body acted on at the point P by a force F at an angle θ to the horizontal. The force can be resolved into a vertical component $F \sin \theta$ and a horizontal component $F \cos \theta$. A vertical force $F \sin \theta$ and a horizontal force $F \cos \theta$ acting simultaneously at P would have precisely the same effect as the force F. This can be verified by drawing a triangle, like the triangle of Fig. 1*a*.

It is often necessary to consider the tendency of forces to rotate bodies. Fig. 1*c* represents a rigid body which is free to rotate about an axis through O, at right angles to the plane of the paper. The force F acts as shown at P, and tends to make the body rotate clockwise about O. Its tendency to do so is measured by its moment about O, Fx, where x is the perpendicular distance between O and the line of action of the force. The force F' acting at P′ tends to rotate the body anticlockwise. If we take the clockwise direction as positive we must regard the moment of F' about O as negative, so it is $-F'x'$.

When two equal parallel forces act on a body in opposite directions, they tend to make it rotate. They are described as a couple. The sum of their moments about any axis perpendicular to the plane in which they act is Fd, where F is the magnitude of each force and d is the perpendicular distance between the lines of action of the two forces.

A rigid body is in equilibrium if the forces acting on it tend neither to move it in any direction nor to rotate it about any axis. Happily, it is not necessary to consider all possible directions and all possible axes to show that a body would be in equilibrium when it was acted on by a particular combination of forces. If the forces all act in the same plane it is sufficient to show that:

(i) The sums of the resolved components of all the forces in each of two directions in the plane are zero. It is often convenient to choose two directions at right angles to each other, but it is not necessary to do so; and (ii) the sum of the moments of all the forces about any one axis at right angles to the plane is zero.

The conditions can also be expressed in other ways, in terms of resolved components in one direction and moments about two axes, or of moments about three axes. These alternative forms of the conditions are explained by Ramsey (1941). If the forces are not confined to a single plane, it is necessary to consider both resolved components in three directions and moments about three axes (for details see Ramsey, 1941).

There are two corollaries to the conditions for equilibrium which are often useful. First and most obvious, a body can only be in equilibrium under the action of two forces if they are equal and act in opposite directions, along the same line. Secondly, if a body is in equilibrium under the action of three forces, the resultant of any two of the forces must be equal and opposite to the third force. This can only be the case when the three forces act in a single plane, and when their lines of action are either parallel or meet in a single point. We will use this corollary in the next section of this book.

Jaws of mammals and their ancestors

The mammals evolved from the synapsid reptiles. As they evolved towards mammalian structure, the synapsids seem to have acquired more and more powerful jaw muscles. The muscles themselves are of course not preserved in the fossils, but the skulls have progressively more space for them and there seems little doubt that the muscles did in fact get bigger. At the same time, the bones at the posterior end of the lower jaw became progressively weaker. This enigma has been explained by considering the equilibrium of the forces acting on the jaws (Crompton, 1963). The explanation depends on the difference in arrangement of the jaw muscles between reptiles and mammals.

Fig. 2a represents the skull of a primitive reptile. The lower jaw is a simple bar with no large processes, and it consists of several bones including the dentary which bears the teeth. The rest of the skull consists of an outer shell of dermal bone with the braincase hidden inside it. The jaw adductor (biting) muscles must have originated in the space between this outer shell and the inner braincase, and inserted on the dorsal edge of the lower jaw. Most of their fibres must have run more or less at right angles to the jaw, as in modern reptiles such as *Sphenodon*

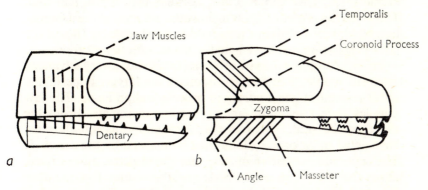

Figure 2. Diagrams showing the arrangement of the jaw muscles in (*a*) a reptile and (*b*) a mammal

(Romer, 1956). Their probable course is indicated in the figure.

Fig. 2b shows the skull and jaw muscles of a mammal. The dentary is the only bone in the lower jaw, and it has quite a complicated shape. It has a dorsal process (the coronoid process) and a sharp posterior ventral corner (the angle of the jaw). There are several distinct jaw muscles acting in different directions. The temporalis muscle runs from the braincase to the coronoid process. It does not run at right angles to the jaw but pulls backwards as well as upwards. The superficial part of the masseter muscle runs from the zygomatic arch to the angle of the jaw, and pulls forwards and upwards. The deeper part of the same muscle (hidden in the figure) runs more vertically. The pterygoideus muscles run from the underside of the skull to the median face of the lower jaw and, like the superficial masseter muscle, pull upwards and forwards. The temporalis, masseter and pterygoideus muscles are all jaw-closing muscles.

Fig. 3 shows the principal forces which probably acted on the lower jaws of various extinct reptiles, when they bit food between their back teeth. Fig. 3a represents the jaw of one of the most primitive reptiles. The jaw muscles exert an upward force (CM) on the jaw. The jaw exerts an upward force on the food and the food exerts an equal downward force (B) on the jaw. If the jaw is stationary, pressing against the food, it must be in equilibrium. It cannot be in equilibrium under the action of the forces CM and B alone, because they are not in line. Therefore the jaw must press against the cranium at the articulation so that the cranium exerts a downward reaction (R) on the jaw. This force must have acted when this reptile bit food.

Fig. 3f represents the jaw of *Diarthrognathus*, a very advanced synapsid reptile. The dentary is much the largest bone in the lower jaw and it has a coronoid process and an angle. The jaw muscles were apparently arranged as in mammals. From the shape of the skull, it seems that the temporalis muscle must have acted more or less along the line marked by the arrow T. The superficial masseter muscle must have acted more or less along the line of the arrow SM, and the resultant of the deep masseter and pterygoideus muscles probably acted more or less along the same line, so T and SM can be taken as representing

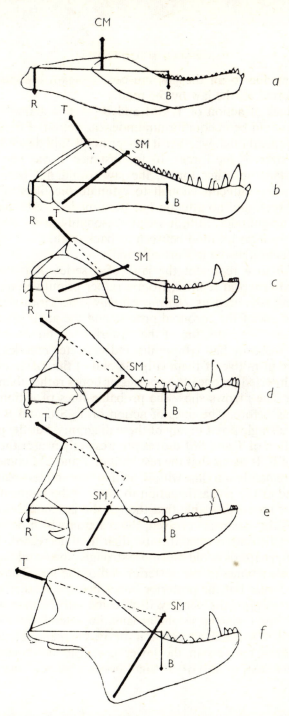

Figure 3. The jaws of a series of fossil reptiles showing the probable lines of action of the forces which acted on them when food was bitten between the back teeth. This figure is explained more fully in the text. (From Crompton, 1963)

the forces due to all the jaw muscles. B again represents the force exerted on the jaw by the food.

The lines of action of T, SM and B meet in a single point. The jaw could be in equilibrium under the action of these three forces alone, though whether it was or not would depend on the relative sizes of the forces. If, for instance T was very much bigger than the other forces, the jaw would not be in equilibrium. It is difficult enough to estimate the lines of action of the forces, and no attempt has been made to establish their relative magnitudes. Still, it seems reasonably likely that when *Diarthrognathus* bit food between its back teeth, there was little or no reaction at the jaw articulation.

Fig. 3*b* to *e* represent the jaws of a series of increasingly advanced synapsid reptiles, from a rather primitive one to one almost as advanced as *Diarthrognathus*. The series shows the development of the coronoid process and the angle of the jaw, and progressive reduction of the posterior jaw bones. All these reptiles probably had temporalis and masseter muscles, but in the more primitive of them (Fig. 3*b* and *c*) the masseter seems to have inserted on one of the posterior bones rather than on the dentary. The arrows show the probable lines of action of the forces, as before. The lines of action of T, SM and B do not meet in a single point in any of these diagrams, but the point of intersection of T and SM moves progressively nearer the line of action of B. It seems that the resultant of T and SM comes more and more nearly into line with B, so that less and less reaction R is needed at the jaw articulation to balance the moments.

It thus seems probable that though the jaw muscles became larger as the synapsids evolved, the reaction which occurred at the jaw articulation when they bit their food decreased. Though the later synapsids could bite more strongly, they could make do with weaker bones at the posterior end of the jaw. The dentary became larger but the posterior bones became smaller until, in *Diarthrognathus*, the dentary itself came into contact with the cranium and formed a second articulation lateral to the original one. Only this second articulation survives in mammals and only the dentary remains in the lower jaw. The other jaw bones have been lost, or survive as small bones with new functions in the ear.

Among the mammals, there are considerable differences in the shape of the lower jaw, and in the relative sizes of the various jaw muscles. These can be correlated with the very different ways in which mammals with different feeding habits use their jaws (Smith and Savage, 1959).

Fig. 4 consists of outlines of the lower jaw of *Martes*, the marten, which feeds on squirrels and other small animals. The arrows represent forces which we will discuss: they show both the lines of action of the forces and (by their lengths) the relative sizes of the forces. In each case, the jaw would be in equilibrium under the action of the forces shown.

Carnivores exert large forces on their jaws in two main ways. First, they use their incisors and large canine teeth to seize their prey or to tear its flesh. When they do this, a force must act on the front of the jaw which has a component forwards (if the animal is tugging at the prey) and a component downwards (since the upper jaw presses on the lower one). This force is represented by the arrow P in Fig. 4*a*, *b*. In Fig. 4*a* it is being resisted by the temporalis muscle alone, which is applying a force T to the coronoid process. The reaction R of the cranium on the jaw balances the forces T and P. Strong bones may be needed at the jaw articulation, but there is no danger of the joint being dislocated because the jaw is being pulled backwards and slightly upwards into its socket. In Fig. 4*b* the force P is being resisted by the masseter muscles alone, which exert the force M on the jaw. Whereas T was reasonably nearly in line with P, M is almost at right angles to it, and a very large force R is needed at the jaw articulation to give equilibrium. This is the case even though M has been deliberately moved forward from its probable natural position (see Fig. 4*d*) to a more favourable position for resisting P. If it had not, R would have been even greater. Notice the direction of R; it would have to be supplied by the ligaments of the jaw articulation, holding the jaw back while M and P tended to pull it out of its socket. If carnivores used only their masseter muscles when they seized prey they would be in danger of dislocating their jaws, but if they used only their temporalis muscles they would not. The temporalis muscles are the largest jaw muscles in carnivores (Table 1).

The masseter is important in the second main use carnivores

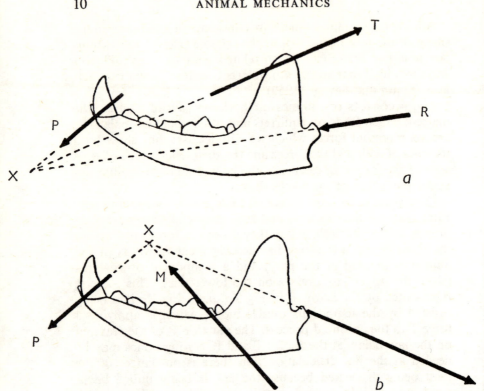

make of their jaws. They use the big carnassial teeth, well back in their jaws, to cut through flesh. This is what a lion or a dog is doing when it sits with a bone or a joint of meat between its paws, gnawing with the corner of its mouth. A roughly vertical force must act on the lower jaw at the carnassial tooth. This force is indicated by the arrows P′ in Figs. 4c, d. Equilibrium could apparently be obtained with little or no reaction at the jaw articulation if the masseter and temporalis muscles contracted together (Fig. 4d). This situation is essentially similar to the one we analysed in *Diarthrognathus* (Fig. 3f). If only the temporalis muscle contracted when the carnassial teeth were used, a considerable reaction at the articulation would be needed for equilibrium (Fig. 4c). The jaw would be pushed backwards and slightly downwards, and there might be some danger of dislocation.

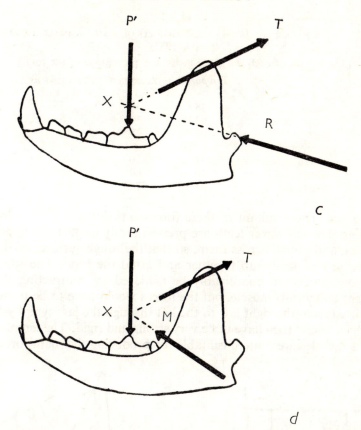

Figure 4. The lower jaw of *Martes*, showing the forces which would act on it in various circumstances which are described in the text. The arrows run along the supposed lines of action of the forces and the lengths of the arrows show the relative sizes of the forces. The jaw would be in equilibrium under the action of each of the combinations of forces which is shown. (From Smith and Savage, 1959)

Herbivores such as horses and cattle use their jaws quite differently. They use their premolar and molar teeth for grinding their food. This involves side-to-side movements of the lower jaw. The jaw is swung to one side, for instance to the left as shown in transverse section in Fig. 5a. The left masseter and pterygoideus muscles contract exerting the forces M and Pt on,

(The weight of each muscle is given as a percentage of the total)

	Temporalis	Masseter	Pterygoideus
Carnivores:			
tiger	48	45	7
bear	64	30	6
dog	67	23	10
Herbivores:			
zebra	11	50	40
European bison	10	60	30
horse	11	57·5	31·5

the jaw. The resultant of these forces acts dorsally and to the right. The left lower teeth are pressed firmly against the upper ones, and pulled across them, so that the rough surfaces of the two sets of teeth rub together and grind the food. The same movement could conceivably be achieved by contracting the right temporalis muscle, but the forces would have to be transmitted from the right jaw to the left through the jaw symphysis which would then have to be very strong and rigid. The temporalis muscles are quite unsuitable for producing the main forces

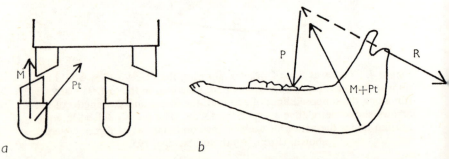

Figure 5. (*a*) A diagrammatic transverse section through the skull of a herbivorous mammal in the region of the cheek teeth, showing how food is ground. The left lower teeth are about to be drawn across the upper teeth by the action of the masseter muscle (which exerts the force *M*) and the pterygoideus muscle (which exerts the force *Pt*). (*b*) Lateral view of the lower jaw of a horse, showing the forces involved in grinding food, as described in the text. The arrows show the lines of action of the forces and, by their lengths, the relative sizes of the forces required for equilibrium

required for grinding. The front teeth are used only for plucking grass, and though this involves forces similar to the force P of Fig. 4a, b they are relatively small forces; there is no need for a large temporalis muscle to prevent dislocation of the jaw. Temporalis muscles are of relatively little use to herbivores, and in them are small. (Table 1.)

Fig. 5b is a side view of the lower jaw of a horse. Its shape is typical of herbivores and very different from the jaws of *Martes* (Fig. 4) and other carnivores. The angle of the jaw, on which the large masseter muscle inserts, is large. The coronoid process, on which the small temporalis muscle inserts, is small. Fig. 4 also shows the forces which are likely to act on the jaw in grinding. There is probably some reaction at the articulation but it may not be as large as appears from the figure because the small temporalis muscle has been entirely ignored and because the force P might not be absolutely vertical. It could have a backward component, due to friction between the upper and lower teeth, which would bring it more nearly into line with the force (M + Pt) due to the masseter and pterygoideus muscles.

Mechanical advantage

A machine is a device for applying forces. One applies a force F_1 to one part of the machine and it applies a force F_2 at another. These forces are not necessarily equal. A small force at the end of a lever can be made to produce a much bigger force near the fulcrum, as can be shown by taking moments. This is the principle of nutcrackers. The ratio F_2/F_1 is known as the mechanical advantage of the machine. The distance the point of application of F_1 has to be moved, to move the point of application of F_2 a unit distance, is known as the velocity ratio of the machine. For a perfect machine, such as a lever with a frictionless fulcrum, the mechanical advantage would be equal to the velocity ratio. For real machines it is always less.

Smith and Savage (1956) compared the mechanical advantages of muscles in the fore-legs of the horse and an armadillo. The skeletons of these legs are shown in Fig. 6. In each case, a broken line from the scapula to the humerus shows the line of action of the teres major, which is one of the main muscles used for swinging the leg back. Two measurements are indicated on

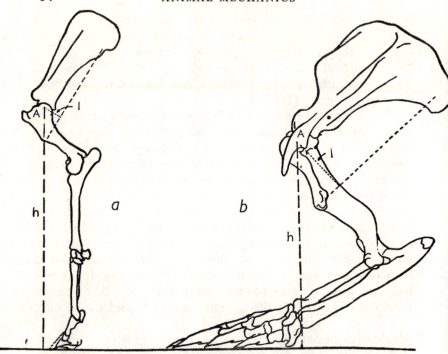

Figure 6. The skeletons of the left fore-legs of (*a*) a horse (*Equus*), and (*b*) an armadillo (*Dasypus*), to show the difference in the mechanical advantage of the teres major muscle. (From Smith and Savage, 1956. Reproduced with permission from *J. Linn. Soc.* (*Zool.*), 42, 1956)

each drawing. h is the distance from the shoulder joint to the ground and l the distance from the shoulder joint to the line of action of the teres major. When the teres major exerts a force F_1 on the humerus, the moment of this force about the shoulder joint is $F_1 l$. It makes the foot exert a force F_2 on the ground whose moment about the shoulder joint is $F_2 h$. If there were no friction at the joint these moments would balance and the mechanical advantage of the teres major, F_2/F_1, would be l/h. In practice, it will be slightly less than this. To move the foot through a distance x, relative to the body, it is only necessary for the teres major to shorten by an amount xl/h. The velocity ratio is l/h.

The horse is adapted for fast running. It moves its legs

very rapidly but does not need to exert particularly large forces on the ground. It needs a low velocity ratio and can tolerate the low mechanical advantage that goes with it: l/h is about 1/13. The armadillo does not run fast, but uses its front legs for digging and so has to exert large forces against the ground. It needs a reasonably high mechanical advantage and can make do with a correspondingly high velocity ratio: l/h is about 1/4.

Fig. 6 also shows that a horse has a relatively smaller olecranon process on its ulna than does the armadillo. The triceps muscle which extends the elbow inserts on this process and has a much smaller mechanical advantage in the horse than in the armadillo.

The horse and the armadillo are not isolated examples. Running mammals in general have fore-leg muscles with low mechanical advantages, and burrowing mammals in general have ones with higher mechanical advantages. Adaptations of mechanical advantage to different habits can also be demonstrated in other groups of animals and other parts of the body.

The mechanical advantage of a muscle is not necessarily constant. If the horse leg in Fig. 6 were swung forwards, the line of action of the teres major would pass nearer the shoulder joint and the mechanical advantage of the muscle would be less. Brown (1963a) estimated the mechanical advantage of the muscle that extends the "knee" of a locust's hind leg. He found that it was 1/60 when the knee was flexed, 1/35 when it was partly extended and 1/60 again when it was fully extended.

Arrangement of fibres in muscles

Voluntary muscles consist of striated muscle fibres, arranged in various ways. There are two main types, known as parallel-fibred and pinnate muscles. They are discussed here because a change in the arrangement of the fibres in a muscle can have an effect equivalent to a change in mechanical advantage.

The two types of muscle are shown schematically in Fig. 7. In the typical parallel-fibred muscle (Fig. 7a), all the fibres run the whole length of the muscle. They are attached at their ends to the skeleton directly or by a tendon (in arthropods, an apodeme), but these attachments are not shown in the diagram. In the typical pinnate muscle (Fig. 7b), the fibres may be much

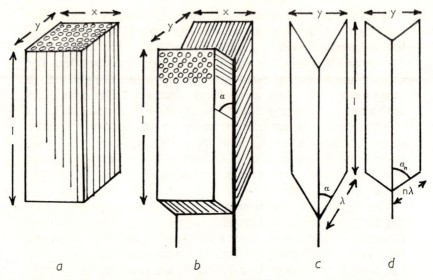

Figure 7. Diagrams illustrating the discussion in the text of parallel-fibred and pinnate muscles

shorter than the muscle and are inclined at an angle to its length. They converge from either side on a central tendon or apodeme to which they are attached.

A long, relatively slender parallel-fibred muscle like the one shown in Fig. 7*a* shortens through a considerable distance when it contracts and exerts a relatively small force. A pinnate muscle of similar shape (Fig. 7*b*) has more, shorter fibres. When it contracts its tendon moves through a shorter distance but can exert more force. Making a long muscle pinnate is equivalent to increasing its mechanical advantage.

Consider first the parallel-fibred muscle (Fig. 7*a*). The resting length of the fibres is l. If they contract to a fraction n of this length, the insertion of the muscle moves a distance $(1 - n)l$. If each fibre has cross-sectional area A when it is relaxed, the muscle contains xy/A fibres. The force which a contracting muscle fibre can develop depends on how far it is allowed to contract (page 109). If it is F_n when the fibre is a fraction n of its resting length, the total force exerted by the contracted muscle is xyF_n/A.

Now consider the pinnate muscle represented in Fig. 7b. It has the same volume as the parallel-fibred muscle and very nearly the same shape, when both are relaxed. Fig. 7c, d shows it relaxed and contracted. When it is relaxed its fibres have length λ and are inclined at an angle α to the tendon, so that

$$\sin \alpha = y/2\lambda$$

When the fibres contract to a length $n\lambda$, they raise the central tendon and the angle between them and the tendon increases to a value α_n where

$$\sin \alpha_n = y/2n\lambda = \sin \alpha/n$$
and
$$\cos \alpha_n = \sqrt{[n^2 - \sin^2 \alpha]}/n$$

The distance the tendon moves is

$$\lambda \cos \alpha - n\lambda \cos \alpha_n$$
$$= \lambda(\cos \alpha - \sqrt{[n^2 - \sin^2 \alpha]})$$
$$= (\cos \alpha - \sqrt{[n^2 - \sin^2 \alpha]})y/2 \sin \alpha$$

The number of fibres in the muscle is $2xl \sin \alpha/A$ since each fibre occupies an area $A/\sin \alpha$ on the face of the muscle, and there are fibres on both sides of the tendon. When they contract, each exerts a force F_n whose resolved component along the tendon is $F_n \cos \alpha_n$. The total force along the tendon is therefore

$$2 xl \sin \alpha . F_n \cos \alpha_n/A$$
$$= 2 xl F_n \sin \alpha \sqrt{[n^2 - \sin^2 \alpha]}/nA$$

The ratio of the force exerted by the pinnate muscle to the force exerted by the parallel fibred muscle is thus $2l \sin \alpha$ $\sqrt{[n^2 - \sin^2 \alpha]}/ny$. This ratio may or may not be greater than 1: the pinnate muscle may exert larger or smaller forces than the parallel-fibred muscle, depending on the values of α, n and l/y.

Few muscles can contract beyond the stage when their fibres are 70% of their resting length (i.e. $n = 0.7$: Gans and Bock, 1965). The force along the tendon becomes zero when $n = \sin \alpha$ because the contracted fibres are then perpendicular to the tendon. This undesirable state will not normally be reached, so long as $\sin \alpha$ is less than 0.7 ($\alpha < 45°$). In fact, α seems usually to be about 30°, or rather less.

Consider a pinnate muscle with $\alpha = 30°$, $\sin \alpha = 0 \cdot 5$. The force it can exert will be $l\sqrt{[n^2 - 0 \cdot 25]}/ny$ times the force exerted by a similar parallel-fibred muscle, contracted to the same extent. At the resting length ($n = 1$) it will exert $0 \cdot 9\, l/y$ times the force exerted by the parallel-fibred muscle. It will only exert more force than the parallel-fibred muscle, if $0 \cdot 9\, l > y$.

Most readers will have learnt or at least been warned that crabs can exert uncomfortably large forces with their chelae. The muscle which closes the chela is pinnate. Let us compare the force it exerts with the force one would expect of a parallel-fibred muscle, occupying the same space.

Fig. 8a shows the structure of a crab chela. The lower claw is fixed and the upper one pivots at O. The chela is opened by a small muscle which inserts on the movable claw at P, and closed by a large muscle inserting at Q. Both these muscles are pinnate, with the muscle fibres converging from either side on an apodeme. The apodemes are stiff sheets of calcified chitin but their attachments to the claw at P and Q are flexible. In the shore crab, *Carcinus*, the length (l) of the chela closing muscle is about twice its average thickness (y). The angle of pinnation is about 30° when the chela is moderately open. From the formula we

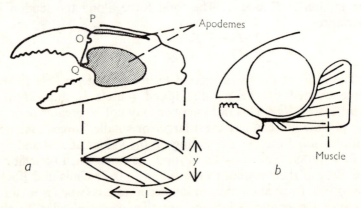

Figure 8. (*a*) The skeleton of a crab chela, opened to show the apodermes and (below) a diagrammatic horizontal section showing the pinnate arrangement of the muscle that closes the chela. (*b*) A diagram showing the pinnate arrangement of the main jaw muscle of a characinoid fish

derived for the idealized pinnate muscle of [Fig. 7b, it seems that the chela closing muscle exerts about twice as much force at Q as it would if it were parallel-fibred. As the function of the chela requires it to nip strongly, it is an advantage for the muscle to be pinnate. However, exactly the same advantage could have been obtained with a parallel-fibred muscle, if its mechanical advantage had been increased by increasing the distance OQ.

Pinnate muscles have another advantage in this and similar situations, where a muscle must contract within a confined space. When a muscle fibre contracts it swells, its volume remaining virtually constant (Baskin and Paolini, 1966). A parallel-fibred muscle swells as it contracts; if it were closely surrounded by rigid walls it could not do so. Pinnate muscles, on the other hand, do not swell. Fig. 7c, d shows why they do not. Contraction does not alter the dimension l, so no swelling is needed to keep the volume constant. The fibres themselves swell but this does not make the muscle swell because of the way they change their angle. The cavity of a crab chela is enclosed by rigid walls and is almost completely filled by the muscles. This is only possible because the muscles are pinnate.

A form of pinnation serves a special function in the jaw muscles of some teleost fish with big eyes, such as *Creatochanes* (Fig. 8b; Alexander, 1964b). The main jaw-closing muscle is crescentic, curving round the ventral and posterior surfaces of the eyeball. If the muscle were parallel-fibred, it would press against the eyeball as it contracted. However, the fibres insert obliquely on a tendon which runs along the concave edge of the muscle, against the eyeball. When they contract, the force each exerts on the tendon has a component acting at right angles to the tendon, pulling it away from the eyeball.

Benninghoff and Rollhäuser (1952) claimed to have shown that a pinnate muscle cannot do as much work in a contraction as a parallel-fibred muscle of equal size. Their proof involved an unfortunate approximation. Had they avoided this approximation, they would have concluded that the two types of muscle do equal amounts of work.

Centre of gravity

In many mechanical problems it is convenient and permissible to think of the mass of a body as being concentrated at a single point, the centre of gravity. We will shortly be analysing the forces acting on a jumping locust. When we do this, we will need to know the position of the locust's centre of gravity. It will be convenient to locate this now, and so illustrate a useful method of finding centres of gravity.

When a body is suspended by a thread, two forces act on it; an upward force due to tension in the thread and a downward force which is the weight of the body. Equilibrium is only possible when these forces act in opposite directions along the same line. The upward force acts along the line of the thread. The weight acts through the centre of gravity. Therefore the suspended body will come to rest with its centre of gravity in line with the thread. The method of finding centres of gravity depends on this.

The position of a locust's centre of gravity depends on the positions of its limbs. When the legs are extended backwards, the centre of gravity moves slightly posteriorly. I therefore used a locust which had been fixed in formalin, in the position we will

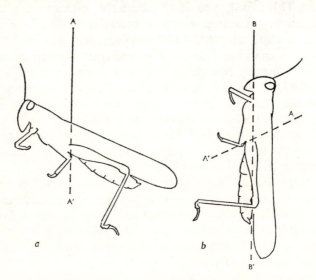

Figure 9. Outlines traced from photographs of a preserved locust, suspended by a thread. The centre of gravity is at the intersection of the lines AA′ and BB′

be considering later in our analysis of forces. I suspended it from a thread attached to its back, and photographed it from the side (Fig. 9a). The centre of gravity must lie on the line of the thread, AA'. I then suspended it from a thread attached to its head and photographed it again (Fig. 9b). The centre of gravity must lie on the line BB'. The centre of gravity must be at the point of intersection of AA' and BB', which can be found by superimposing the two photographs. Since the locust is symmetrical, the centre of gravity must lie in the median plane.

Work and energy

Work is done when the point of application of a force is moved. The work is the distance moved multiplied by the resolved component of the force in the direction of motion. When a force of 1 dyn moves its point of application 1 cm along its line of action, it does 1 erg of work. 1 joule is 10^7 erg. Power is the rate of doing work and is measured in erg/s or in watts (J/sec). If a force of 1 dyn moves its point of application at a velocity of 1 cm/sec along its line of action, the power is 1 erg/s.

Energy is a measure of capacity to do work. A weight held above the ground is said to possess potential energy because it could be used to do work if it were allowed to fall to the ground. It could, for instance, be made to raise another weight attached to it by a rope running over a pulley. A body of mass m gm held h cm above the ground could do mgh ergs of work as it fell, so its potential energy is mgh ergs. A moving body is said to possess kinetic energy because it can be made to do work against a force that resists its motion. The kinetic energy of a body of mass m gm travelling at v cm/s is $\frac{1}{2}mv^2$ erg. This is the amount of work it would do when it was brought to rest. Heat energy can be made to do work, for instance in a steam engine; 1 calorie is equivalent to $4\cdot2$ J. Electrical energy, chemical energy and so on can also be made to do work.

The work which is done is equal to the energy used in doing it. This is the Principle of the Conservation of Energy. However, it is not normally possible to convert all the energy which is used into *useful* work. There is often wastage of energy due to friction between the parts of a machine. A proportion of the work which is done is done against friction, and the energy used

2—AM

in doing this is lost as heat. If a quantity of energy E_1 must be supplied to a machine to do a quantity of useful work E_2, the efficiency of the machine is E_2/E_1. This ratio is often expressed as a percentage.

Efficiency of muscular effort

When we use our muscles we release chemical energy by metabolism and use a proportion of it to do mechanical work, while a proportion is lost as heat. Dickinson (1929) measured the efficiency of conversion of chemical energy to mechanical work, by a person working a device known as a bicycle ergometer.

The subject of the experiment works a pair of bicycle pedals which make a wheel revolve. A strip of cloth running over the wheel acts as a brake. One end of the strip is attached to a weight and the other to a spring balance, as shown in Fig. 10. If the weight has mass m a force mg will act on its end of the band. A smaller force F, measured by the spring balance, acts on the other end. The brake thus exerts a frictional force $(mg - F)$ on the edge of the wheel. If the wheel has radius r and makes n revolutions in unit time, the velocity of its edge is $2\pi rn$. The power required to move the edge of the wheel at this

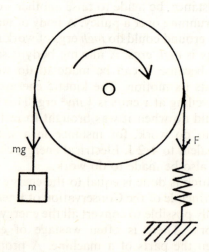

Figure 10. A diagram illustrating the principle of the bicycle ergometer

velocity against the frictional force is $2\pi rn \ (mg - F)$, and can be calculated. Though pedalling against a brake may seem a rather futile activity, this power is the rate of doing "useful work" in the sense intended in the definition of efficiency.

The rate at which chemical energy is being used can be measured indirectly by collecting and analysing the air which the subject breathes out. The method is explained in textbooks of physiology (for instance, Bell, Davidson and Scarborough, 1965). About 5 cal of chemical energy is released for every millilitre of oxygen used in respiration. The exact value depends on the proportions of fat and carbohydrate which are being used, but the rate at which chemical energy is being released can be calculated quite accurately if the carbon dioxide content of the expired air is determined as well as the oxygen content.

Dickinson measured the rate at which her subjects used chemical energy, when they were resting and when they were pedalling the bicycle ergometer. The difference in each case gave the rate at which chemical energy had to be released, to obtain the mechanical power which was being used in the ergometer.

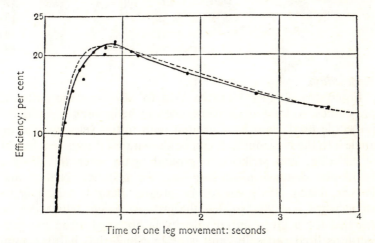

Figure 11. Graph of the efficiency of conversion of chemical energy to mechanical work, by a person pedalling a bicycle ergometer, against the time taken for each leg movement (i.e. for each half revolution of the pedals). The broken line is a theoretical curve, not mentioned in the text.
(From Dickinson, 1929)

Hence she was able to calculate the efficiency of conversion of chemical energy to mechanical power. She found that it varied with the rate of pedalling, as shown in Fig. 11. It had a maximum value of 22% when each leg movement took 0·9 s (i.e. when the pedals revolved once in 1·8 s).

The bicycle ergometer can be used to measure the maximum power that the leg muscles can produce, as well as their efficiency. Parry (1949) calculated from Dickinson's results that the power was up to 40 W/kg muscle. This power can only be exerted for a short time because oxygen cannot be supplied to the muscles fast enough to maintain it.

Motion with constant acceleration

There is a very useful group of equations which describe the motion of a body which has constant acceleration a. Let the component of its velocity in the direction of the acceleration be u at time 0 and v at time t. Let it travel a distance s in the direction of the acceleration between these times. It can easily be shown that

$$v = u + at \tag{2}$$
$$s = ut + \tfrac{1}{2}at^2 \tag{3}$$
$$s = (u + v)t/2 \tag{4}$$
$$v^2 = u^2 + 2as \tag{5}$$

Projectiles

Equations (2) to (5) can be used to work out the paths of projectiles, so long as air resistance can be ignored. Consider a projectile fired from level ground with initial velocity u at an angle α to the horizontal. It can be shown that it rises to a height $u^2 \sin^2 \alpha / 2g$, and strikes the ground again after travelling a horizontal distance $u^2 \sin 2\alpha / g$ (see, for instance, Nelkon and Parker, 1965). Sin α cannot be greater than 1, which is its value when $\alpha = 90°$, so that the greatest height which can be reached for a given initial velocity is obtained when the projectile is fired vertically, and is $u^2/2g$. Sin 2α has its maximum value of 1 when $\alpha = 45°$, so the greatest range for a given initial velocity is obtained when the projectile is fired at 45°. It is u^2/g.

Locust jumping

Locusts (*Schistocerca*) can jump distances up to about 80 cm on a level floor. We can use the formulae we have just discussed to estimate the velocity at which they must leave the ground. If they took off at the optimum angle of 45° the required velocity would be given by the equation

$$u^2/g = 80$$
$$u \quad = \sqrt{[80 \times 981]} = 280 \text{ cm/s}$$

In fact, locusts seem usually to take off at at least 55° (Hoyle, 1955; Brown, 1963a). It would be a little unsatisfactory to jump at so steep an angle if it were required to jump as far as possible, but locusts often use a jump as a means of getting airborne. They jump and, while in the air, start flying. A locust jumping at 55° and travelling 80 cm would have to leave the ground at a velocity given by the equation

$$u^2 \sin 110°/g = 80$$
$$u = \sqrt{[80 \times 981/0 \cdot 94]} = 290 \text{ cm/s}$$

Even this estimate of take-off velocity must be too low, for we have ignored air resistance. Brown (1963a) took high speed cinematograph films of locusts jumping, and measured the velocity directly. He quotes as typical a jump in which a locust left the ground at 340 cm/s.

Locusts use their big hind legs for jumping. Each leg consists of a series of sections, with movable joints between them (Fig. 12*a*). The short section known as the trochanter has a movable joint with the femur in many insects but is rigidly fixed to it in locusts. We shall be looking at the structure of the joints in Chapter 2 (page 37).

The continuous outline in Fig. 12*a* shows the position of the leg before a jump (Brown, 1963a). The tarsus rests on the ground and the "knee" joint between the femur and tibia is bent. When the locust jumps, it accelerates its body to the take-off velocity by extending the hind legs rapidly. This moves the locust's coxa through a distance of about 4 cm to the position indicated by the broken outline before the tarsus leaves the ground. If the velocity at take-off is 340 cm/s, the locust's body must be accelerated from rest to 340 cm/s over a distance of 4 cm. The

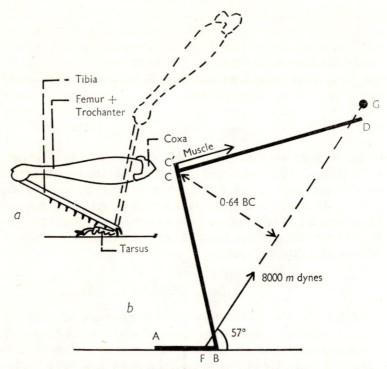

Figure 12. (*a*) A diagram of the hind leg of a locust showing its position before the start of a jump and (broken outline) at the moment when the feet leave the ground. (*b*) The diagram used in the analysis of the forces which act on a locust leg during take-off

acceleration which would be required if it were constant, can be calculated from equation (5).

$$(340)^2 = 2a \times 4$$
$$a = 14{,}500 \text{ cm/s}^2$$

If the mass of the locust is m g, a force of 14,500 m dyn would be needed to give it this acceleration.

It will be convenient to consider a locust taking off at 55°, like the specimen in the sequence of cinematograph photographs published by Brown (1963a). The hind legs must exert the force of 14,500 m dyn on the ground, at an angle of 55° to the horizontal. They must also exert a downward force mg dyn to support the weight of the locust. On page 2, we added these two forces together and found that their resultant is a force of 15,300 m dyn, acting at 57° to the horizontal.

In calculating this, we assumed that the acceleration (and so the force) was constant. Brown (1963a) made a locust jump off a platform fitted with instruments for measuring forces. He found that the force exerted by the locust on the platform was not constant during extension of the legs, but rose to a peak and then declined. The maximum force was about 17,000 m dyn. We will assume in the analysis that follows that each hind leg exerts a force of 8,000 m dyn on the ground, at an angle of 57° to the horizontal.

In Fig. 12b, ABCD represents the hind leg of a locust, at a particular moment during take-off. AB represents the tarsus, BC′ the tibia and CD the femur and trochanter. The position is taken from one of the photographs in the sequence published by Brown (1963a). We have already located the centre of gravity of a locust in this position (Fig. 9, page 20), and it is indicated by the point G. The leg exerts a force of 8,000 m dyn at 57° on the ground, and the ground exerts an equal, opposite force on the leg. It must act along the line FG since the resultant of it and the similar force on the other leg must pass through the centre of gravity if the locust is not to be sent spinning.

Consider the equilibrium of the lower half of the leg, ABC′. It is acted on by the reaction of the ground acting along FG, by the force exerted by a muscle which inserts at C′ and runs along the femur, and by the reaction at the joint C. It is not of course strictly in equilibrium because the tibia is accelerating, but the mass of the tibia is so small compared with the mass of the locust that assuming it is in equilibrium will cause very little error.

We can calculate the force exerted by the muscle by taking moments about the knee joint, C. The muscle acts at a distance CC′ from the joint and according to Brown (1963a) this equals BC/35. The perpendicular distance from FG to C is 0·64 BC. Hence, the force exerted by the muscle must be 8,000 m × 0·64 BC/(BC/35) = 1·8 × 10^5 m dyn. For a 3 g locust, this would be about 5 × 10^5 dyn or 500 g wgt. As Brown remarks, the apodeme of the muscle must be very strong.

We could calculate the forces exerted by the muscles which work the other joints in the same way, but it seems more interesting to compare the work done by the various muscles.

The work done by a muscle is the product of the force F it exerts and the amount d that it shortens. The moment of the force about the joint is Fl if it acts at a distance l from the joint. The angle through which it moves the joint is d/l rad (about $57d/l°$; see page 33). Hence, the work done by a muscle equals the moment about the joint multiplied by the angle in radians through which the joint moves. The reaction of the ground exerts a very much larger moment about the locust's knee joint C than about the joints B and D or the joint between the coxa and the thorax which is not shown in the figure, because FG is so much further from C than from the other joints. We have only demonstrated this for one stage in the extension of the leg but diagrams like Fig. 12*b* could be drawn for other stages and would show the same thing, except at the final stages of take-off when the leg is almost straight. Brown's photographs show the angle BCD between the tibia and the femur increasing by 95° as the leg is straightened, while the angle ABC between the tarsus and the tibia and the angle between the femur and the body, change by 30° and 50° (the published photographs are not clear enough to show whether the coxa moves). Thus, there is a very much larger moment about the knee joint than about the other joints through most of the process of extension of the leg, and the knee joint extends through a larger angle than the other joints. Nearly all the work involved in jumping must be done by the extensor muscle of the knee joint. This is what one would expect on anatomical grounds, for it is very much the largest of the leg muscles.

Parry and Brown (1959b) made an analysis similar to this one of jumps by a jumping spider. Hall-Craggs (1965) analysed a jump by a bushbaby and Badoux (1965) analysed kangaroo jumps.

Jumps by animals of different sizes

Many animals jump, and have adaptations (such as big muscular hind legs) to help them jump high or far. The heights and distances which some of them can jump are shown in

TABLE 2

THE HEIGHTS AND DISTANCES WHICH SOME ANIMALS CAN JUMP

(The acceleration distance is the quantity *l* referred to in the text. Records marked with an asterisk probably refer to running jumps, but the rest refer to standing jumps)

	Acceleration distance (cm)	High jump (cm)	Long jump (cm)	
Man		120*		Dyson, 1962
Man			370	Hill, 1950
Kangaroo (*Macropus*)	100	270*		Hill, 1950
Rat kangaroo (*Bettongia*)		240*		Hill, 1950
Bushbaby (*Galago*)	16	226		Hall-Craggs, 1965
Jumping mouse (*Zapus*)			370	Hill, 1950
Frog (*Rana*)	10		90	Gray, 1953
Locust (*Schistocerca*)	4		80	
Locust (*Schistocerca*)		30		Hoyle, 1955
Unidentified grasshopper	2		75	
Flea (*Spilopsyllus*)	0·04	6		Bennet-Clark and Lucey, 1967

Table 2. As far as possible, running jumps have been excluded from the table. In a running long jump, kinetic energy developed in the run is used to help the jump. The faster the animal is running when it jumps, the further it will travel. Kinetic energy from the run can also, in certain circumstances, help a high jump. This is the principle of pole vaulting (Dyson, 1962).

Man is not particularly well adapted for jumping, but has nevertheless been included in the table. The value of 120 cm given as his high jump record needs explanation. Good athletes can, of course, jump over bars which are much more than 120 cm above the ground, but they do not raise their centres of gravity more than about 120 cm. They start with the centre of gravity about 90 cm above the ground, but it passes very close over the bar. When they use the Western Roll technique, for instance, the centre of gravity may rise only about 15 cm above the bar (Dyson, 1962). The record for the kangaroo should perhaps be adjusted for the same reasons, though the correction would not be nearly as large as for man since the kangaroo which is reported to have jumped a 9 ft (270 cm) fence presumably did not execute a Western Roll. No correction is necessary for the other animals, whose jumps are large compared with their height. The bushbaby which jumped to a height of 226 cm, for instance, must have raised its centre of gravity by very nearly this amount.

Some of the records in the table suggest that the height or distance an animal can jump bears very little relation to its size. A rat kangaroo, which is about the size of a rabbit, can jump almost as high as a large kangaroo. A grasshopper can jump about as far as a locust of twice its length. Is this remarkable? Let us consider some factors which might affect an animal's jumping ability.

When an animal makes a standing jump, each of the muscles which is used contracts once. Jumping ability might depend on the amount of work a muscle can do in a single contraction. The force a muscle can develop is proportional to its cross-sectional area. The amount it can shorten is proportional to its length. Therefore the work it can do in a single contraction is proportional to the product of cross-sectional area and length, which is the volume of the muscle. Equal volumes (or weights) of muscle should be able to do equal amounts of work. Consider an animal of mass m with jumping muscles of mass m'. Let these muscles be able to do an amount of work, km', in a single contraction. The work done by these muscles is equal to the kinetic energy with which the animal leaves the ground.

$$\tfrac{1}{2}mu^2 = km'$$
$$u^2 \quad = 2km'/m$$

where u is the velocity at take-off. If the animal jumped vertically, it would rise to a height $u^2/2g = km'/mg$ (see page 24). If it jumped at 45°, it would travel a distance $u^2/g = 2km'/mg$. One might therefore expect animals with equal proportions of jumping muscles in their bodies (equal values of m'/m) to be able to jump equal heights and distances, irrespective of their size. A rat kangaroo is reasonably similar in shape to a large kangaroo and cannot have a very different value of m'/m, so one might expect it to be able to jump about as high as a large kangaroo. One might expect a grasshopper to jump about as far as a locust.

We will now try the effect of a different assumption about the muscles. We will assume that the jumping ability is limited by the maximum power the muscles can produce, and that unit mass of muscle can produce power k'. Let the animal's centre of gravity travel a distance l between the moment when the muscles start to contract and the moment when the feet leave the ground. For most animals, l will be a little less than the length of the legs. We will use the same other symbols as before. We already know that work $\frac{1}{2} mu^2$ must be done as the animal takes off. To obtain the power we must divide this work by the time t in which it is done. The animal accelerates from rest to velocity u, while moving through distance l in time t. We will suppose that the acceleration is constant, and apply equation (4) (page 24). This gives

$$l = (0 + u)t/2$$
$$t = 2l/u$$

The power involved in doing work $\frac{1}{2} mu^2$ in this time is $mu^3/4l$. The power the jumping muscles can produce is $k'm'$. Hence

$$k'm' = mu^3/4l$$
$$u = (4k'lm'/m)^{1/3}$$

If an animal takes off vertically at this velocity it will reach a height $u^2/2g = (4k'lm'/m)^{2/3}/2g$. If it takes off at 45° it will travel a distance $u^2/g = (4k'lm'/m)^{2/3}/g$. Animals of different sizes, with similar proportions of jumping muscle, should be able to jump heights and distances proportional to the 2/3 power of the acceleration distance l.

Some estimates of l are given in Table 2. If jumps were proportional to $l^{2/3}$, one would expect a kangaroo to jump about $(100/16)^{2/3} = 3 \cdot 4$ times as high as a bushbaby. One would expect a locust to jump $2^{2/3} = 1 \cdot 6$ times as far as a grasshopper. One would be hopelessly wrong. It seems that some small animals produce more power per unit weight of muscle in a jump than similar large animals.

We would not be justified in concluding that the jumping ability of animals is limited solely by the work their muscles can do. For one thing, the conclusion that animals of similar proportions and different sizes should jump equally well could have been reached by supposing that the jump was limited by the strength of the skeleton, using information which will be given in Chapter 4. For another, the similarity between the heights jumped by mammals of various sizes, and the similarity between the distances jumped by the locust and the grasshopper, cannot conceal the enormous difference between the jumping ability of the mammals on the one hand, and the insects on the other.

We will compare the jumps made by the bushbaby and the locust. The bushbaby which jumped to 226 cm (Table 2) must have had a take-off velocity, u, of 667 cm/s. Locusts are known to take off at about 340 cm/s. I have weighed the extensor muscles of a bushbaby's hind legs and find that they total about 9% of the body weight, so m'/m for the bushbaby is 0·09. Hoyle (1955) found that the extensor muscles of the knee joints of the hind legs of a locust totalled about 5% of the body weight and we have seen that these muscles do nearly all the work of jumping (page 28), so m'/m for the locust is 0·05. These data enable us to calculate the work k ($= u^2m/2m'$) done by unit weight of jumping muscles in a jump. It is 0·25 J/g for the bushbaby and 0·12 J/g for the locust. The acceleration distance l is 16 cm for the bushbaby and 4 cm for the locust, so we can calculate the power k' ($= u^3m/4m'l$) exerted per unit weight of jumping muscle. It is 5·2 W/g for the bushbaby and 4·9 W/g for the locust.

The muscles of the bushbaby do about twice as much work per unit weight in a jump as the muscles of the locust, but they seem to exert almost exactly the same power. It would be nice to draw a simple conclusion but it would not be reasonable to do so, for the bushbaby is a warm-blooded animal and the locust is not. Both the work a muscle can do in a contraction and the power it can exert depend on temperature. Locust flight muscles can do 3 times as much work in a contraction at 35°C as at 11°C (Weis-Fogh, 1956b). It has been estimated that frog muscle could exert 10 times as much power at 40°C as at 10°C (Bainbridge, 1961).

The power exerted in a jump is very high. It is about 5 W/g muscle in bushbabies and locusts. It is probably much the same in fleas: a calculation based on Bennet-Clark and Lucey's (1967) data and on their assumption that m'/m cannot be more than 0·2, gives a power of at least 3 W/g. These estimates are a great deal higher than the value of 40 W/kg (0·04 W/g) calculated from the results of experiments with a bicycle ergometer (page 24) but this is partly due to their being values for a single contraction instead of average values over a series of cycles of contraction and relaxation. It has been estimated that frog muscle could develop 0·4 W/g in a single contraction at 40°C (Bainbridge, 1961) but this is still very well below the values estimated from jumps. It seems that the high powers exerted in jumping, at least by locusts and fleas, are only made possible by use of the principle of the catapult (see page 80).

Circular motion

If a body is travelling in a circle, there are several ways of saying how fast it is going. We may give its velocity, v. We may state how many revolutions it makes in unit time. We may give its angular velocity, which is the angle it travels through in unit time. This is usually measured in radians per second.

A radian is the angle subtended at the centre of a circle by an arc whose length equals the radius. Since the circumference of a circle of radius r is $2\pi r$, a radian is $1/2\pi$ of a revolution or about 57°. A body making n revolutions every second has an angular velocity of $2\pi n$ rad/s. A body travelling with angular velocity ω in a circle of radius r has velocity $r\omega$.

The velocity is constant in magnitude but not in direction. Therefore, a force is needed to keep the body travelling in a circle. It must act towards the centre of the circle and is known as centripetal force. If it is not applied, the body will fly off at a tangent. If the mass of the body is m, the force which is needed is $m\omega^2 r$ ($= mv^2/r$). We will use this expression when we consider the forces which act on birds and gliders, when they turn (page 243).

Moments of inertia

The kinetic energy of a body of mass m moving in a straight line with velocity v is $\frac{1}{2} mv^2$. The kinetic energy of a rotating body is given by a similar expression, $\frac{1}{2} I\omega^2$, where I is the moment of inertia of the body about its axis of rotation and ω is its angular velocity.

Moment of inertia is best explained by showing how it is determined. We will use as an example an insect wing. What is the kinetic energy of the wing as it swings up or down with angular velocity ω, when the insect flies?

The data needed to calculate the moment of inertia of an insect wing can be obtained by cutting it into strips, making each cut across the wing at a measured distance from its base, and weighing the strips (Sotavalta, 1952). Fig. 13 shows a wing with a single strip marked on it. The axis which is marked is the axis about which the wing rotates as it swings up and down. Different parts of the strip are different distances from this axis but the differences are small since the strip is narrow. We will not introduce much error if we measure the distance r from the axis to the centre of the strip and regard the whole mass of the strip as being concentrated at this distance from the axis.

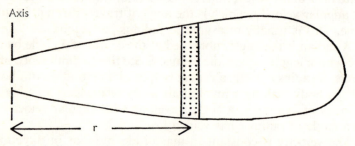

Figure 13. A diagram of an insect wing illustrating the account of moments of inertia

If the mass of the strip is m and its angular velocity about the axis ω, its kinetic energy will be very nearly the same as that of a body of mass m moving in a straight line with velocity $r\omega$. It will be $\frac{1}{2}mr^2\omega^2$. The kinetic energy of the whole wing will be the sum of the values for all the strips, $\sum(\frac{1}{2}mr^2\omega^2)$. Since the angular velocity of an intact wing is the same for all its parts, ω is the same for all the strips and we can re-write the expression for the kinetic energy $\frac{1}{2}\omega^2\sum(mr^2)$. Compare this with the expression $\frac{1}{2}I\omega^2$, given at the beginning of this section. I, the moment of inertia of the wing about the axis marked in the figure, is given by $\sum(mr^2)$, the sum of the values of mr^2 for all the individual strips. If the values of m are measured in grams and the values of r in centimetres, the moment of inertia is $\sum(mr^2)$ g . cm².

When one gives a moment of inertia, one must specify the axis of rotation. The wing shown in Fig. 13 would have lower moments of inertia about an axis half-way along it, and about an axis running from its base to its tip, than about the axis shown.

Later in this book we will discuss the energy needed for insect flight (page 264). The wings have a certain kinetic energy in the upstroke, and a certain kinetic energy as they move in the opposite direction in the downstroke. At the end of each stroke the wings must be stopped (i.e. kinetic energy must be taken from them) and accelerated in the opposite direction (i.e. kinetic energy must be supplied to them). To estimate the amounts of energy involved we must know the moments of inertia of the wings.

Moments of inertia also figure in discussions of running and galloping (Smith and Savage, 1956). A large proportion of the energy used by a mammal running on level ground is used in accelerating the limbs, alternately forwards and backwards. The lower the moments of inertia of the limbs, the less energy is needed. The moments of inertia will be lower if the muscle is mostly around the hip and shoulder, than if it extends far down the legs, because moment of inertia depends on distance from the axis of rotation. Fast mammals have plenty of meat in the proximal halves of their limbs, and very little in the distal halves.

CHAPTER 2

JOINTS AND MECHANISMS

THE branches of mechanics which form the basis of this chapter are dealt with in textbooks for engineering students, on the theory of machines (for instance Bevan, 1956). It will be necessary to introduce some concepts which will be unfamiliar to most zoologists, but which are conveniently simple.

Kinematics and joints

All possible movements of a rigid body can be described in terms of three axes, mutually at right angles. Any linear movement can be described in terms of its resolved components along the three axes. Any rotation can similarly be described in terms of components about the three axes. To describe a position of the body fully, one must state six quantities. A body which can be moved in any direction and revolved about any axis is therefore said to have six degrees of freedom of movement (Bulleid, 1922).

A rigid body which is movably attached to others is known as a link. If it is attached to another whose position is regarded as fixed, it has less than six degrees of freedom. For instance, a hinge joint would only allow it to rotate one axis, and would not allow any linear movements. Two links joined by a hinge joint have only one degree of freedom of movement relative to each other. There are two other simple types of joint which allow only one degree of freedom of relative movement: they are sliding joints, which only allow sliding along one axis, and screw joints such as exist between a bolt and its nut. A screw joint allows movement along an axis and rotation around it, but these are strictly proportional to each other so only one quantity is needed to specify a change of position and only one degree of freedom is involved. A link joined to a fixed link by a ball and socket joint can rotate about any axis through the centre of the ball, and since all its rotations can be described in terms of components about three axes has three degrees of freedom of movement.

Arthropod joints

Many insect larvae have flexible cuticle but in most adult arthropods the cuticle forms a stiff exoskeleton. Movements are made possible by joints, where two regions of thick exoskeleton are separated by a flexible region. There are examples of the main types of joint in insect legs, such as the hind leg of a locust or grasshopper (Fig. 12a; Snodgrass, 1935).

The joints between the subsegments of the tarsus in the locust are of the simplest type. Each consists merely of a band of flexible cuticle. There is no contact between successive pieces of stiff cuticle. Such a joint allows a wide variety of small movements of which some are illustrated in Fig. 14a. Indeed, such a joint allows rotation about any axis and linear movement in any direction. Two links connected by such a joint have six degrees of freedom of relative movement, just as if they were not connected. Such a joint limits the extent of the movements that are possible but does not reduce their variety.

The structure of the joint between the coxa and the thorax varies among insects, but it is usually a ball and socket joint. It

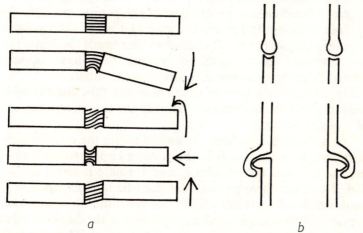

a   b

Figure 14. Diagrams of arthropod joints which are described in the text

must be admitted that it is often more complicated than the simple joint which will be described. There is a flexible articular membrane between the thorax and the coxa but in addition there is a knob on the thorax which fits a hollow in the wall of the coxa. The knob and hollow are held firmly together by the nearby part of the articular membrane. The coxa can rotate about any axis relative to the thorax.

The trochanter and femur usually form a single rigid unit. The coxa is joined to the trochanter, and the femur to the tibia, by a dicondylic joint. A knob articulates with a hollow on each side of the limb and the only possible movement is rotation about an axis through the two knobs. In other words, these dicondylic joints serve as hinges. Two types of dicondylic joint are shown in section in Fig. 14b. In the upper example the two knobs point in the same direction along the limb but in the lower example they point laterally in opposite directions. The upper example is only held together by the articular membranes but the lower one cannot be dislocated except by forces large enough to break the skeleton. The femoro-tibial joints of insects are like this, and so are the joints in the legs of crabs.

Different joints require different numbers of muscles. Very often, one finds a pair of muscles for each degree of freedom of relative movement allowed by the joint. For instance, the hinge joint between the femur and the tibia in an insect leg allows one degree of freedom and has one pair of antagonistic muscles. One of these muscles straightens the joint and is the main jumping muscle in locusts (page 27) while the other bends the joint. They are described as antagonistic because they have opposite effects. When one contracts, it stretches the other.

Fig. 15 shows the lines of action of most of the muscles which work the ball and socket joint between the thorax and the coxa of a typical insect.

The antagonistic muscles M and N rotate the coxa in either direction about the axis b–b, the muscles I and J rotate it about the axis c–c and the muscles K and L rotate it about the axis d–d. There are three pairs of antagonistic muscles operating a joint which allows three degrees of freedom. However, there are typically two more muscles, not shown in the diagram, which insert on the ledges on either side of the socket. They can assist

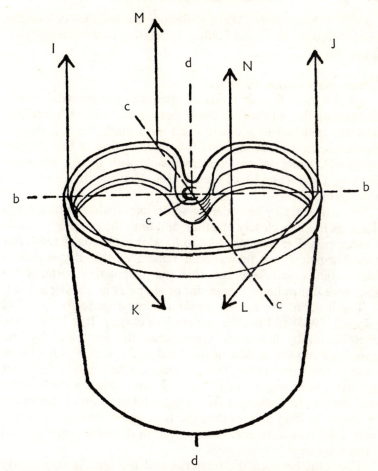

Figure 15. A diagram of the proximal end of an insect coxa, showing the lines of action of muscles which insert on it. (From Snodgrass, *Principles of Insect Morphology.* Copyright 1935 by permission of the McGraw-Hill Book Co)

M by contracting together and they can assist I or J by contracting separately, but they add nothing to the variety of movements possible at the joint. The coxae of the middle legs of bees have dicondylic joints with the thorax which allow only one degree of freedom. They have four muscles, which is less than in insects with ball and socket joints but more than is necessary to operate a hinge joint. We will discuss the number of muscles needed to operate a joint further, in relation to mammal elbows (page 46).

Kinematics of mammal joints

All the more mobile joints of mammals are synovial joints: the articulating surfaces of the bones are covered with cartilage and are enclosed in a cavity filled with fluid. We will discuss the lubrication of these joints in a later section of this chapter (page 61). Here, we are concerned only with their kinematics. Barnett, Davies and MacConnaill (1961) have written a book on synovial joints.

We will start by considering some joints whose articulating surfaces fit fairly closely together, and limit the range of possible movements. Readers may like to look at skeletons as they read this discussion, but they should remember that dry skeletons lack articular cartilage and that the articulating surfaces are therefore not quite the same size or shape as in a living animal.

The head of the human femur is almost perfectly spherical (Rydell, 1966) and fits the acetabulum closely. It is held in the acetabulum by ligaments surrounding the joint and by one running from the middle of the head of the femur to the lower edge of the acetabulum. The hip joint is a ball and socket joint allowing three degrees of freedom. I can swing my leg forward and back as in walking, I can swing it laterally and I can rotate it about its long axis so that the toes point forward or to the side. Different ligaments become taut in different positions and limit the range of movement.

The joint between the humerus and the ulna is very nearly a simple hinge joint. The only considerable movement it allows is rotation about the axis indicated in Fig. 16a. This movement occurs when the elbow is bent and extended. If the articulating surfaces were cylindrical and there were no ligaments to prevent it, the ulna could also slide from side to side, parallel to the axis. In fact, the articulating surface of the humerus is grooved like a pulley wheel and the ulna is shaped to fit it (Fig. 16a), so sliding is prevented. The whole elbow is enclosed in a capsule of collagen fibres but parts of the capsule are thickened to form distinct ligaments. Fig. 16b shows how the

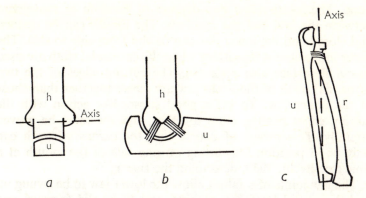

Figure 16. Diagrams of the bones and some of the ligaments of a human
elbow and forearm: (*a*) is a section, and (*b*) is a medial view
h, humerus; *r*, radius; *u*, ulna

ligaments are arranged on the medial side of the elbow. They
radiate from attachments on the humerus, close to the axis of
the hinge. This radiating arrangement means that the distance
between the attachments of the two ends of each ligament
scarcely changes, as the joint moves. Both ligaments are more
or less taut at all positions of the joint. They neither become
slack as the joint moves, nor check its movements. In this they
contrast with the ligaments of the hip. There is another radiating
ligament on the lateral side of the elbow joint, from the
humerus to the annular ligament which will be described in the
next paragraph.

The radius can move relative to the ulna so that when the
forearm is horizontal the palm of the hand can be turned up
(the supine position) or down (the prone position). The radius
rotates about the axis indicated in Fig. 16*c*. At the end nearest
the elbow it has a short cylindrical head which rotates against
an articular facet on the ulna. The radius is held against the ulna
by an annular ligament attached at each end to the ulna. The
ligament encircles the radius without being attached to it. The
other end of the radius has a concave articular surface which
articulates with a convex surface on the ulna.

The lower jaw of a cat has roughly cylindrical articular
condyles, which fit into transverse grooves on the squamosal

bone. The joints allow two degrees of freedom of movement between the jaw and the cranium. The mouth can be opened and closed and the lower jaw can be slid from side to side. The latter movement is important when the carnassial teeth are used to cut flesh (see also page 10). The cutting edges of the two carnassial teeth of the lower jaw are closer together than those of the upper jaw. To get a good scissor-like action with the cutting edges passing close together, the jaw must be moved laterally. This means of course that the carnassial teeth can only be in position for cutting on one side of the mouth at a time (see Becht, 1953, describing the tiger).

The jaw joints of a rabbit allow the lower jaw to be swung up and down and from side to side, and to be slid forward and back. All these movements are important in feeding (Ardran, Kemp and Ride, 1958). It would be impossible to devise a single joint with closely fitting surfaces which would allow this range of movements. The articulating surfaces do not fit each other at all closely.

There are many other joints whose articulating surfaces do not fit closely. In the human knee, for instance, the convex condyles on the lower end of the femur rest on the almost flat upper end of the tibia (Fig. 17). The bones are held together by various

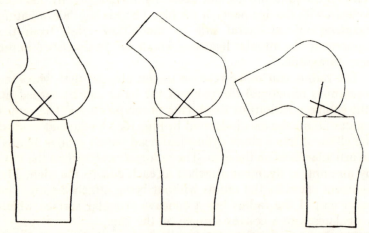

Figure 17. Diagrams of a human knee joint, showing the action of the cruciate ligaments. (From Barnett, Davies and MacConnaill, 1961)

ligaments, including the strong cruciate ligaments. These run in the gap between the two condyles of the femur and cross, as the figure indicates. The one with the anterior attachment to the tibia is the anterior cruciate ligament, and the other is the posterior cruciate ligament. The figure shows how they affect the movements of the joint. When the knee bends, the posterior cruciate ligament pulls the femur posteriorly relative to the tibia. When it extends, the anterior cruciate ligament pulls it forward again (Barnett, Davies and MacConnaill, 1961). Because the articular surfaces do not fit closely, only small areas of the articular cartilages are in contact at any one time. The arrangement of the cruciate ligaments ensures that these areas change as the joint moves. When we examine the theory of weeping lubrication of mammal joints (page 62) we will see that these points may be important. Joints like the knee joint, where the articulating surfaces do not fit closely and sliding occurs as well as rotation, often have wedge-shaped fibrocartilages which partly fill the gaps where the bones fail to meet (Barnett, 1954).

If the shapes of the articulating surfaces and the positions of attachment of the cruciate ligaments were suitably adjusted, the knee could be bent and extended without either slackening the cruciate ligaments or stretching them. This situation is represented in Fig. 17; if the ligaments are measured, each will be found to have the same length in all three diagrams. In fact the situation is slightly different. The shapes of the articulating surfaces are such that the ligaments are stretched as the knee extends, and so help to limit extension of the joint.

We can learn something about the numbers of muscles required to operate joints, by studying the muscles which move our forearms. Let us examine the muscles which move the ulna relative to the humerus, and the radius relative to the ulna. The joints have already been described.

The triceps is the main muscle extending the elbow. This muscle runs from the scapula and back of the humerus to the olecranon process of the ulna. The triceps is assisted by the much smaller anconeus. The brachialis runs from the front of the humerus to the ulna, and bends the elbow. The remaining muscles are shown in Fig. 18. The supinator runs from the humerus and ulna to the radius, and pulls the radius to the supine position. The pronator teres and pronator quadratus pull the radius to the prone position. The biceps runs from the scapula to the radius. It both bends the elbow and pulls the radius to the supine position. The brachioradialis runs from the humerus to the distal end of the radius. It bends the elbow and pulls the radius to a position intermediate between the prone and supine positions.

There are so many of these muscles that most movements of the joints could be performed in several ways, using different

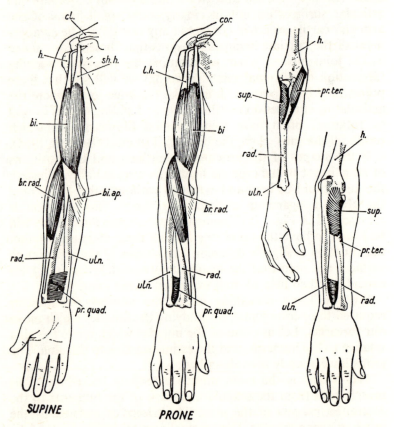

SUPINE   **PRONE**

Figure 18. Simplified drawings showing some of the muscles which move the human forearm. (From Young, 1957)
bi, biceps; *br. rad.*, brachioradialis; *h*, humerus; *pr. quad.*, pronator quadratus; *pr. ter.*, pronator teres; *rad.*, radius; *sup.*, supinator; *uln.*, ulna

muscles or groups of muscles. The muscles actually used to perform some of the movements have been identified by inserting needle-like electrodes into muscles, and recording their electrical activity (Basmajian, 1962). When action potentials are recorded from a muscle, it must be contracting.

The brachialis muscle is active whenever the elbow is bent. The biceps is also used in bending the elbow when the forearm is in the supine position, but not when it is in the prone position. If it were used in the latter case, another muscle would have to be used at the same time to prevent it from supinating the forearm. When the forearm is supinated with the elbow held extended, as in tightening a screw with the right hand, the biceps is not used unless considerable force is needed. When it is used, the triceps or anconeus must be used as well to prevent it from bending the elbow. Ordinary supination is done by the supinator alone.

The brachioradialis has little or no activity when the elbow is simply being held bent so long as no great force is being resisted, but is very active when the elbow is bent rapidly. All the muscles that bend the elbow have some activity when it is being extended, but the brachioradialis is particularly active when the elbow is extended rapidly. Since it is very active whenever the forearm is rotating rapidly about the elbow joint, in whichever direction, it is believed that one of its functions is to provide centripetal force and so protect the elbow from dislocation. The biceps and brachialis cannot do this because they run along the humerus and are only in line with the forearm when the elbow is extended. The brachioradialis runs along the forearm and can provide centripetal force at all positions of the elbow. The biceps and brachialis muscles could bend the elbow just as strongly if they, too, ran along the forearm, but if they did the forearm would be unwieldy, as they would increase its moment of inertia about the elbow joint (see page 35). If the brachioradialis is able to provide the centripetal forces that are required, it is best to have the rest of the elbow muscles proximal to the joint. The brachioradialis may have another function, which is suggested on page 139. It may protect the bones of the forearm from breakage, by keeping the stresses in them reasonably low.

All possible movements of the elbow and forearm could be performed by the biceps, the triceps and the pronator. The elbow could be extended by the triceps. It could be bent by the biceps and pronator, acting together (the pronator would prevent the biceps from supinating the forearm). The forearm could be pronated by the pronator, and supinated by the biceps and triceps acting together (the triceps would prevent the biceps from bending the forearm). Although three muscles could perform all the movements, there are actually eight. One reason for having more than three muscles is probably that it is uneconomic, for instance, to use the biceps and triceps together to supinate the forearm. The triceps would do no mechanical work in the process since its insertion would not move, but chemical energy would be used in maintaining its tension. We have seen that the supinator is used in preference to the biceps when supination without bending is required, and that the brachialis is used without the biceps when bending without supination is required. Though all the movements of the forearm could be produced by three muscles, they could be produced more economically by four (one each for flexion, extension, pronation and supination). The brachioradialis seems to have a special function in providing centripetal force, so we can explain the presence of five muscles. The biceps originates on the scapula and is used in movements of the shoulder as well as in movements of the elbow. With more knowledge, we might be able to explain the presence of all eight muscles.

The joints between the humerus and the ulna, and between the ulna and the radius, are hinge joints allowing one degree of freedom each. The forearm thus has two degrees of freedom of movement relative to the humerus. We have seen that all the movements it can make could be produced by three muscles. More generally, it can be shown that a system of bones and joints allowing n degrees of freedom of relative movement can be operated by $(n + 1)$ muscles. More muscles than this minimum are very often found, for reasons such as the ones we have been discussing. The minimum can be reduced to n if one muscle is replaced by elastic material: for instance the shell of the scallop, *Pecten*, is closed by a muscle but opened by elastic recoil of the hinge ligament (page 82).

Kinematic chains

A kinematic chain is an assembly of links jointed together in such a way that if one is fixed and another moved, all the rest must move in a predictable manner. The parts of a kinematic chain have one degree of freedom of movement relative to each other. Fig. 19*a* shows a very simple kinematic chain, described as a four-bar crank chain. It consists of four links joined by hinge joints whose axes are parallel. If one of the links is fixed and another moved, the movements of the rest are predictable. If the angle between any two links is given, the rest of the angles can be worked out.

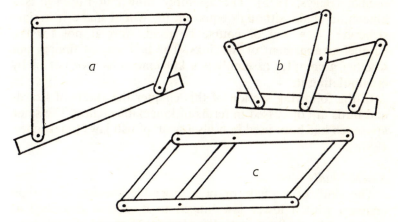

Figure 19. Three kinematic chains

A loop of three links hinged together would not be a kinematic chain, but a rigid structure: the links could not be moved relative to each other. A loop of five links hinged together would not be a kinematic chain but would have two degrees of freedom of relative movement. If one link was fixed and another moved, the rest could make a variety of movements. It would be necessary to give two angles by specify a position.

Simple kinematic chains can be joined together to make more complicated ones. The arrangement shown in Fig. 19*b* is an example. It consists of two four-bar crank chains which have two links in common. If one of the six links is fixed and another is moved, the rest will move in a predictable manner.

The number of degrees of freedom of relative movement X of an assembly of L links joined by J joints can be predicted by means of equations (Bottema, 1950; Beggs, 1955). There are various equations, applying to assemblies with different types of joints. If all the joints are hinges whose axes are parallel

$$X = 3(L - 1) - 2J \qquad (6)$$

When $X = 1$, the assembly is a kinematic chain. When $X \leqslant 0$, it is a rigid structure.

Though the equations generally hold, it is regrettably easy to design assemblies which have more degrees of freedom than they predict (Beggs, 1955). The assembly shown in Fig. 19c is a kinematic chain although equation (6) would make it a rigid structure. It would of course be rigid, were it not for the parallel arrangement of the links. The best way of finding out whether a complex assembly is a kinematic chain or not, is by manipulating it.

In the following section of this chapter the skulls of lizards and birds are discussed in terms of kinematic chains. The same approach has been used in a discussion of fish jaws (Alexander, 1967a).

Kinetic skulls

The skull of a turtle or of a crocodile consists of a rigid cranium with a lower jaw hinged to it, and is described as akinetic. The skulls of many lizards, of snakes and of birds have movable joints in the cranium and are described as kinetic. The next few paragraphs are about the mechanisms of kinetic skulls.

Frazzetta (1962) investigated the skulls of lizards and treated them largely as problems in engineering. Fig. 20a is a diagram representing the main parts of a typical lizard skull. The suture between the frontal and parietal bones is straight and transverse, and forms a hinge between the muzzle and the posterior part of the skull roof. The quadrates are hinged to the posterior part of the skull roof. The pterygoids are firmly but flexibly attached to the quadrates by ligaments, and are sutured to the palatines at their anterior ends. The palatines are incorporated in the muzzle but they and the pterygoids are flexible, so we can think of the pterygoids as being hinged to the muzzle, as shown

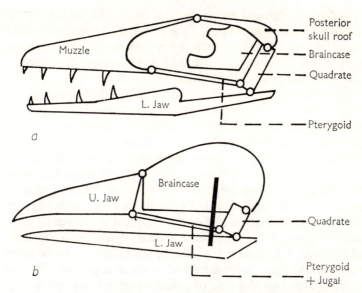

Figure 20. Diagrams showing the mechanisms of the kinetic skulls of (*a*) a lizard, and (*b*) a bird. The ligament from the braincase to the lower jaw of the bird is shown black

in the diagram. The ectopterygoids and jugals are attached to the pterygoids and move as a unit with them but their attachment to the muzzle is flexible and acts as a lateral extension of the pterygoid-muzzle joint. The braincase can be moved slightly relative to the overlying parts of the skull roof and it has processes which contact the pterygoids. Its attachment to the pterygoids is loose and does not affect the jaw mechanism, so we need not consider the braincase further. The two halves of the lower jaw are joined firmly together at the symphysis and are hinged to the quadrates.

The muzzle, the posterior part of the skull roof, the pterygoids and the quadrates form a four-bar crank chain like that of Fig. 19*a*. The quadrates and pterygoids are of course pairs of bones, not single ones, but this does not affect the mechanism. If, for instance, the posterior part of the skull roof is held stationary and the quadrates are moved, the pterygoids and muzzle must move in a particular way. If the ventral ends of the quadrates are pushed forwards, the pterygoids move forward and the muzzle swings up. If they are pushed back, the muzzle swings down. These movements are independent of the movements of the lower jaw, so the skull has two degrees of freedom of relative movement of its parts.

Cinematograph films of lizards feeding show the use they make of skull movements. When they open their mouths to seize prey, the muzzle swings up and the lower jaw swings down. When they close their mouths on their prey, the muzzle swings down as the lower jaw swings up. This may help the lizard to get a firm grip quickly. If only the lower jaw moved, it would strike the prey before the muzzle and would have to raise the prey to get a firm grip with both sets of teeth. Lizards often bite their prey several times to kill it. A typical sequence of movements is shown in Fig. 21. The muzzle is swung up in Fig. 21*b, h* and down in Fig. 21*d, j*. The muzzle seems always to be swung up as the lower jaw is lowered and down as it is raised. Though the skull has two degrees of freedom of relative movement, they do not seem to be exploited fully. Indeed, the main jaw-closing muscle is arranged in such a way that it tends to lower the muzzle as well as raising the jaw.

The skulls of snakes have many more movable joints than the skulls of lizards, and a correspondingly large number of degrees of freedom of relative movement. Snakes make a remarkable variety of skull movements when they feed on large prey (Frazzetta, 1966).

The skulls of birds are much more like those of lizards, from a mechanical point of view. Fig. 20*b* is a diagram representing the skull of a typical bird (Bock, 1964). A flexible lamina of bone which connects the braincase to the upper jaw acts as a hinge joint. The dorsal part of the roof of the skull is incorporated in the braincase. The quadrates are hinged to the braincase and flexibly attached to the pterygoids. The anterior ends of the pterygoids are attached to the palatines which are flexible and serve as hinges between them and the upper jaw. The jugals lie lateral to the pterygoids and have similar attachments. They share the mechanical function of the pterygoids and are not shown separately in the diagram. The two halves of the lower jaw are fixed rigidly together and are hinged to the

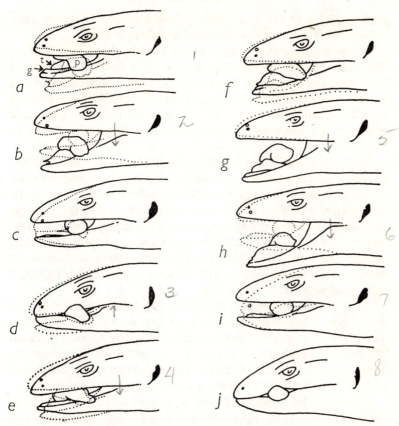

Figure 21. Outlines traced from a sequence of cine photographs of a lizard (*Gerrhonotus*) chewing prey. The dotted outlines show the next position in the sequence. (From Fig. 7, T. H. Frazzetta, *J. Morph.* 111: 287)

g, gums; *p*, prey; *t*, tongue

quadrates. A ligament runs from the braincase, just behind the eye, to the lower jaw. It is made of collagen and is virtually inextensible. It is known as the postorbital ligament.

The upper jaw, braincase, quadrates and pterygoids form a four-bar crank chain. If the braincase is held and the quadrates are swung forwards, the upper jaw is raised. In this, the bird resembles the lizard. However, the movements of the upper and lower jaws are not entirely independent. They are only independent when the postorbital ligament is slack.

When the ligament is taut, it has exactly the same effect on the skull mechanism as if it were a rigid link, hinged to the braincase and lower jaw. The ligament, braincase, quadrates and lower jaw then behave as a second four-bar crank chain, sharing two links with the first. The whole skull forms a kinematic chain of six links and seven joints, like the chain shown in Fig. 19b.

Consider a bird's head with the beak closed. If the lower jaw is pulled down it tightens the postorbital ligament and the skull behaves as a kinematic chain. Because the ligament cannot be stretched the posterior end of the lower jaw swings up, pushing the quadrate forward and up and raising the upper jaw. Lowering the lower jaw automatically raises the upper jaw. However, raising the upper jaw does not force the lower jaw down. It merely slackens the ligament and allows the lower jaw to fall, if it is otherwise free to do so.

Now consider a bird's head with the beak wide open. The lower jaw is lowered, and the upper one is raised. If the lower jaw is raised it does not force the upper jaw down, but it slackens the ligament and allows the upper jaw to fall. If, however, the upper jaw is lowered it makes the quadrate swing back and down, tightening the ligament and raising the lower jaw.

The postorbital ligament makes it easier for the bird to keep its mouth shut. If the upper jaw is heavy enough, the mouth will remain closed when all the jaw muscles are relaxed, because the lower jaw cannot sink without raising the upper one.

The thorax of a fly

Engineers have to devise mechanisms which will produce particular movements. Zoologists are often faced with the reverse problem, of explaining the mechanism that gives rise to an observed movement. This is sometimes best done by means of working models.

When a fly flies its wings move in a complicated way. As they swing up and down, they rotate about their long axes. In the downstroke the anterior edge of each wing is a little lower than the posterior edge but in the upstroke the plane of the wing becomes nearly vertical, with the anterior edge above the posterior edge (Fig. 111, page 261). We will see in Chapter 6 how the wing movements produce the forces needed for flight. In this chapter we will simply examine the mechanism that produces the movement.

Dipteran flies have only one pair of wings. The thorax that bears them is a complicated box whose walls are stiff in some places and flexible in others. The way it works is best explained by means of the working model devised by Pringle (1957). It is not necessary for us to examine the actual structure of a fly: it is sufficient to understand the model, each part of which represents a part of the thorax. Any reader who cares to look up the actual anatomy of the thorax (Boettiger and Furshpan, 1952, summarized in Pringle's book) will soon see how much the model clarifies the mechanism.

The model is shown in Fig. 22. The parts on the left represent the anterior parts of the thorax. The four largest components of the model form a four-bar crank chain (Fig. 23a, b) whose movements are controlled by two Bowden cables which represent a pair of antagonistic muscles. Bowden cables are flexible tubes with wire down the middle, such as are used to operate the brakes of some bicycles. When the "muscle" dvm contracts the four-bar chain is pulled into the position shown in Fig. 23a. When the "muscle" dlm contracts the top component is pulled posteriorly and the chain adopts the position shown in Fig. 23b. Note that z (the end of the scutellar lever) rises when dlm contracts, and falls when dvm contracts.

Fig. 23c, d are transverse sections showing some of the other parts of the model. The piece of elastic, psm, represents a pair of muscles which pull the side walls of the thorax towards the mid-line. The wings pivot on the side walls of the thorax at x and are attached to the scutellar lever (z in Fig. 23a, b) at y. When the muscle dlm contracts and raises z, the wings swing down. When dvm contracts and lowers z, the wings swing up.

Fig. 22b shows the attachments of the wing in more detail. There are ball and socket joints at x, y and f. A peg protruding from the socket of the joint at y fits loosely into a rectangular hole in the end of z. This arrangement keeps the two points y a constant distance apart. Hence, when the wings move up or

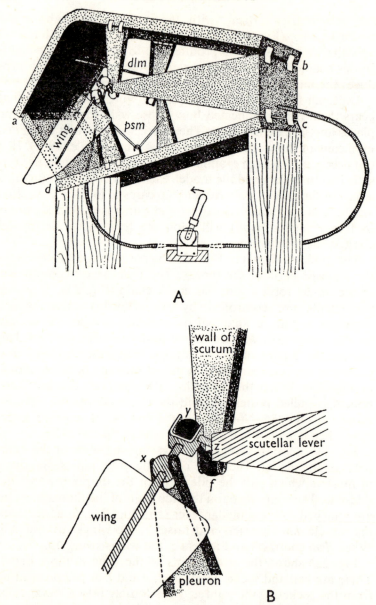

Figure 22. A model of a fly's thorax to show the mechanism of the wing
movements (From Pringle, 1957)

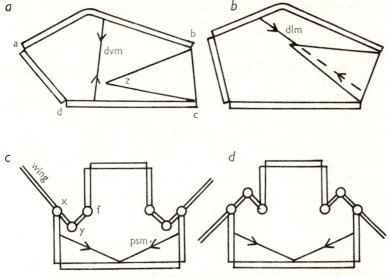

Figure 23. Diagrams of the model shown in Fig. 22, with the wings raised (A, C) and lowered (B, D)

down, between the positions of Fig. 23*c* and *d*, the points *x* are pushed laterally at first, and then move together again when the wings have passed the horizontal position. The lateral movements of *x* are resisted by the "muscle" *psm* so the wings have a "click" action: the first half of each wing stroke is slow, resisted by *psm* and the second half is fast, aided by it.

When the wings are raised, the model is in the position indicated by Fig. 23*a*, *c*. When *dlm* contracts, *z* rises. It does not at first swing the wings down, because this is resisted by *psm*. The peg is pushed up but *y* is not, and the wing rotates about its long axis so that its anterior edge is lower than its posterior edge. Once this has happened, further upward movement of *z* can only raise *y* and swing the wing down. The same sort of thing happens in the upstroke. When *dvm* contracts, *z* swings down. At first it merely rotates the wing about its long axis, so that the anterior edge of the wing is higher than the posterior edge. Only when this has happened does the wing start swinging up, against the resistance of *psm*. Thus, alternate contractions of two muscles and constant tension in another suffice to produce the rather complicated cycle of wing movements.

Friction

Friction is dealt with in this chapter, because of its importance in relation to joints.

A brick resting on a table does not move when it is pushed, unless the force exceeds a certain minimum, equal to the maximum possible (limiting) frictional force. Frictional forces act parallel to the surfaces which are in contact but the limiting frictional force between two surfaces is proportional to the force at right angles to the surfaces, pressing them together. If F_{max} is the limiting frictional force and N is the normal force which each surface exerts on the other

$$F_{max} = \mu N \qquad (7)$$

where μ is a constant for the surfaces in question, known as the coefficient of static friction. The limiting frictional force seems independent of the area of contact between two surfaces: it is the same for a brick set flat on a table as for a brick set on end. However, no surface is really smooth and it seems that the real area of contact between two surfaces is very small and is proportional to the normal force. Equation (7) does not hold exactly for surfaces made of plastics and other high polymers (Ritchie, 1965).

When one surface is moving over another, a frictional force resists its motion, but it is less than the limiting frictional force for stationary surfaces. Less force is needed to keep a brick moving over a table, than to start it moving. The ratio of the frictional force when the surfaces are moving with constant velocity, to the normal force, is known as the coefficient of dynamic friction. It is nearly independent of the velocity. Energy used in doing work against friction is converted to heat.

Coefficients of friction between metals, between plastics or between a metal and a plastic, are usually between $0 \cdot 15$ and $0 \cdot 8$ (Hodgman, 1965; Ritchie, 1965).

Locking fin spines

In most joints friction is a nuisance, and we will see in the next section of this chapter that it is lessened by a remarkable system of lubrication in the synovial joints of mammals. This section deals with a joint in which friction is exploited for a special purpose.

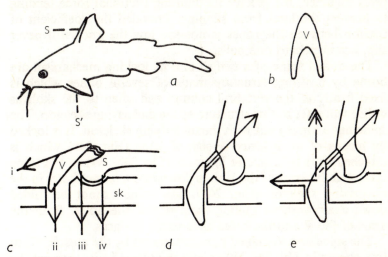

Figure 24. A typical catfish and diagrams illustrating the account in the text of the mechanism of the dorsal spine
S, dorsal spine; *S'*, pectoral spine; *sk*, underlying skeleton; *V*, V-shaped bone; *i–iv*, muscles

Sticklebacks, catfishes (Fig. 24a) and some other fish have strong sharp spines in their fins which seem to discourage predators from eating them. The spines can be folded flat against the body but when danger threatens they are raised so as to project at right angles from it. No predator can then eat the fish, without being wounded by the spines. The system would be less effective if predators could fold the spines down again by pressing on them with their jaws. Sticklebacks and catfishes do not rely on muscular strength to keep the spines erect, for the spines are locked erect by friction and cannot be folded down by pressing on their tips. We will examine the locking mechanism of the single spine in the dorsal fin of catfishes (Alexander, 1966a). Different mechanisms based on the same principle lock the pectoral fin spines of catfishes and the spines of sticklebacks.

The principle is essentially that of knots. The harder one pulls on two ropes which are knotted together, the more tightly they press together at the knot (i.e. the greater the normal forces at the points of contact between the ropes). The more tightly they press together, the greater the limiting frictional force tending to prevent the knot from slipping. Provided the coefficient of friction between the ropes is not too low the knot will never slip, however hard one pulls.

The dorsal spine of a catfish and its locking mechanism are borne by a strong structure made of several bones which is fixed firmly to the vertebral column and often to the skull as well. I will refer to this structure as the underlying skeleton. The spine has a hinge joint with the underlying skeleton. It is locked by means of an inverted V-shaped bone (Fig. 24b) which is attached to it by a strong ligament (Fig. 24c). When the spine is erected the arms of the V slide down into a pair of holes in the underlying skeleton (Fig. 24d). The spine cannot be folded down again without pulling them out of the holes. Both the arms of the V and the holes have rough surfaces.

The spine and V-shaped bone are moved by the muscles (i) to (iv), shown in Fig. 24c. When muscles (ii) and (iii) contract the spine is erected and the arms of the V-shaped bone are pulled down into the holes. If the tip of the spine is then pushed backwards, the spine pulls on the V-shaped bone through the ligament (Fig. 24d). This does not extract the V-shaped bone from the holes but pulls it crooked, jamming the rough surfaces together. The spine cannot be depressed, simply by pressing its tip. It can only be depressed if a more or less vertical force acts on the V-shaped bone and pulls it out of the holes instead of jamming it. The catfish applies this force to depress the spine, by contraction of muscles (i) and (iv). Muscle (i) pulls anteriorly on the V-shaped bone. Muscle (iv) pulls obliquely upwards on it, through the ligament. The resultant of the two forces acts roughly vertically (Fig. 24e). The V-shaped bone is pulled out of the holes and the spine folds down.

Let us consider the conditions necessary for the locking mechanism to work. Fig. 25 shows a simplified model in which the V-shaped bone is represented by the rod, AB, which fits into a hole of slightly greater diameter than itself. We will

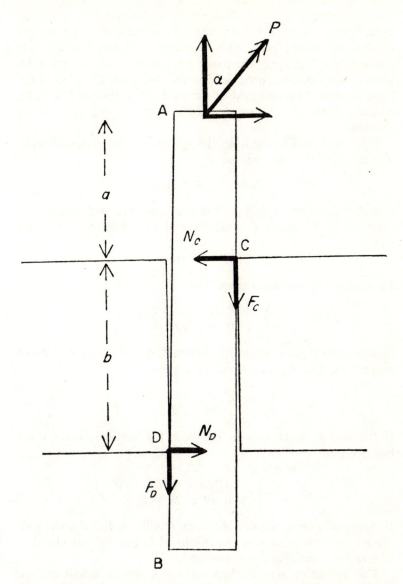

Figure 25. This diagram is explained in the text. (From Alexander, 1966a)

ignore the weight of the rod as the weight of the V-shaped bone is small compared to the other forces which act on it. Consider what happens when a force P acts on the rod at A at an angle α to the axis of the hole. This represents the force which acts through the ligament when one tries to depress the spine by pushing its tip. The force deflects the rod from the axis of the hole, so that it presses against the wall of the hole at C and D. Normal forces N_C, N_D and frictional forces F_C, F_D then act on the rod.

If the rod jams in the hole, the system will be in equilibrium. The horizontal forces will balance.

$$N_C = N_D + P \sin \alpha$$

and the moments about C will balance. We will ignore the moment of F_D about C which would be relatively small and write

$$bN_D = aP \sin \alpha$$

From these two equations we find

$$N_C = P \sin \alpha . (a + b)/b$$
$$N_D = P \sin \alpha . a/b$$

If the coefficient of static friction at C and D is μ, F_C cannot exceed μN_C and F_D cannot exceed μN_D, so

$$F_C \leqslant \mu P \sin \alpha . (a + b)/b$$
$$F_D \leqslant \mu P \sin \alpha . a/b$$

If the rod is to remain in the hole the vertical forces must balance and

$$P \cos \alpha = F_C + F_D$$
$$\leqslant \mu P \sin \alpha . (2a + b)/b$$
$$\mu \geqslant b \cot \alpha/(2a + b)$$

If this condition is satisfied the force P will not dislodge the rod. No matter how great it is, the frictional forces will withstand it. The rod will remain firmly jammed.

This condition was worked out for a simple model but we may apply it cautiously to a real fish. I estimate very roughly that in the catfish *Pimelodus* $\alpha = 45°$ and $b/(2a + b) = 0 \cdot 3$.

This gives as the condition for jamming

$$\mu \geqslant 0 \cdot 23$$

which is a very modest requirement. I have not measured the coefficient of friction, but the locking mechanism of *Pimelodus* is certainly effective.

Lubrication of mammal joints

When I walk, I do work against friction in my hip joints. The radius of the head of my femur must be about 2 cm. As I take a step, the leg that is supporting my weight rotates through about a radian (57°) at the hip, so the articular surface of the femur must slide about 2 cm over the surface of the acetabulum. The downward force on the head of the femur is greater than the weight of the body, because the centre of gravity of the body is not immediately above the joint, so that muscles lateral to the joint have to contract to balance the moments. The force is 1·6 or more times the weight of the body, depending on the speed of walking (Rydell, 1966). If the mass of my body is m g and the coefficient of dynamic friction at the joint μ, the frictional force will be at least $1 \cdot 6 \, \mu m g$ dyn. The work done against the frictional force is obtained by multiplying the force by the distance; it is about $3 \, \mu m g$ erg. This is equal to the work required to lift my whole body $3 \, \mu$ cm. It would be quite an appreciable amount of work if the coefficient of friction lay in the usual range for unlubricated solid surfaces. Not only would unlubricated joints make extra work for the muscles, but they would also be liable to overheating. The energy used doing work against friction would be released in the joints as heat. It is possible that vigorous activity might heat the articulating surfaces to a temperature which caused damage. Proteins are easily damaged by heat.

Friction can be greatly reduced by lubrication. Coefficients of dynamic friction well below $0 \cdot 01$ are obtained at lubricated bearings in machines (Bevan, 1956), compared with values of around $0 \cdot 2$ between dry metals. When a stationary shaft rests in its bearing the oil is squeezed out from under it, but when it rotates the oil is dragged round with it owing to viscosity. A thin layer of oil separates the metal surfaces so long as the shaft rotates. This is hydrodynamic lubrication. It is only effective when there is fairly fast continuous movement in the same direction. It is ineffective when movement is slow, or when one surface moves backwards and forwards over the other. It does not seem to have any importance in animal joints.

Once a metal surface has been oiled, an exceedingly thin film of oil remains on it which is difficult to remove. Friction between such surfaces is less than between clean metal surfaces, though more than in hydrodynamic lubrication. The phenomenon is called boundary lubrication. It is effective even when the surfaces move slowly, or backwards and forwards, but it does not seem to give low enough coefficients of friction to explain the performance of mammal joints.

The coefficients of friction in mammal joints have been measured by using a joint as the pivot for a pendulum. When the pendulum is set swinging its swings get gradually smaller. This is partly due to air resistance, but largely to friction in the joint. If the rate at which the swings get smaller is observed, the coefficient of friction can be calculated. Values of about 0·01 have been obtained (Charnley, 1959; Barnett and Cobbold, 1962).

We must now examine the structure of a synovial joint; that is, of any of the more mobile joints in a mammal. All the joints in our limbs are synovial joints, but the joints between successive vertebrae are not. Fig. 26 represents a typical synovial joint. The end of each bone is covered by a layer of cartilage. The joint is enclosed by a membrane and the cavity is filled with synovial fluid, the properties of which will be described later (page 86). The cartilage is like a sponge with very fine pores and it contains synovial fluid which can be squeezed out of it. Lewis and McCutchen (1959: McCutchen 1959, 1962a, b) suggested that this sponginess might make possible a system of lubrication, unknown in engineering, which they called weeping lubrication.

They demonstrated weeping lubrication in experiments with closed-cell sponge rubber. This is the sort of sponge rubber that contains individual bubbles of air rather than a network of channels like the pores of a bath sponge. They cut the sponge rubber to expose a layer of cavities, and measured the coefficient of friction between the cut surface and glass, with

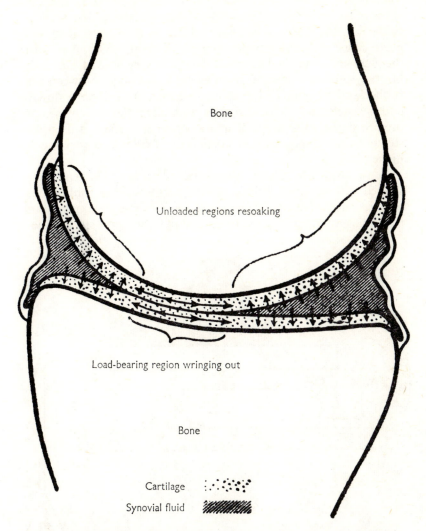

Figure 26. A diagrammatic section through a synovial joint, illustrating the theory of weeping lubrication. (From McCutchen, 1962b)

soapy water as a lubricant. They found it was only about 0·003. The reason seems to be this: the cavities exposed at the cut surface are filled with soapy water, and nearly all the load is borne by this water when the rubber is pressed against the glass. Since the walls of the cavities which are in contact with the glass are so easily deformed, they bear very little of the load, but they prevent the water from escaping sideways, between the rubber and the glass. The trapped water seeps out very slowly and the coefficient of friction rises gradually if the sponge is kept pressed against the glass. If a plain rubber surface is used instead of the spongy one, the coefficient of friction is very much higher.

Lewis and McCutchen suggested that articular cartilage soaked with synovial fluid behaves like the sponge rubber and soapy water. The cartilage is not a closed-cell sponge but resembles a bath sponge with exceedingly fine pores. The fine-

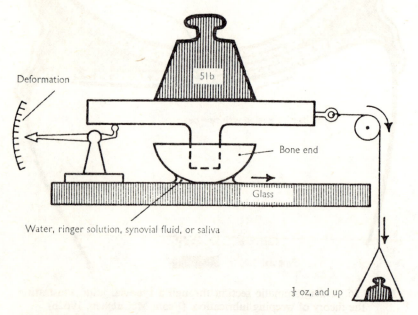

Figure 27. Apparatus for measuring the coefficient of friction between the articular surface of a bone and glass, and the amount of deformation (squashing) of the articular cartilage. (From McCutchen, 1962b)

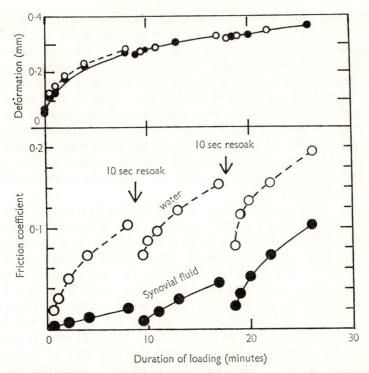

Figure 28. The results of experiments with the apparatus of Fig. 27. The duration of loading is the time since the 5 lb weight was first set in position. The weight was removed for 10 s at each of the times indicated. The surfaces were lubricated with water or synovial fluid, as indicated. (Redrawn from McCutchen, 1962b)

ness of the pores prevents the synovial fluid from being squeezed out too quickly. They investigated the possibility, using the method indicated in Fig. 27 to measure the coefficient of friction between the head of a pig's humerus and glass. They used glass instead of the scapula, to simplify the design of the experiment. The apparatus was set up and two quantities were measured at intervals. The horizontal force needed to make the humerus slide was measured so that the coefficient of friction could be calculated. The distance between the glass and the plate on which the humerus was mounted was measured to find out how much the articular cartilage had been squashed. Some results are shown in Fig. 28. The continuous lines show results obtained with the cartilage soaked in synovial fluid, as in the living animal. When the apparatus was first set up, the coefficient of friction was about 0·003. As time passed the cartilage got squashed (indicating that synovial fluid was being squeezed out of it) and the coefficient of friction rose. These changes suggest weeping lubrication. So does the effect of removing the load for 10 s. There is a very slight reduction in the deformation, and the coefficient of friction is reduced temporarily when the load is replaced. Apparently the 10 s allowed the cartilage to soak up again a little of the fluid it had lost. Complete recovery takes about half an hour.

Fig. 26 shows what is thought to happen in a typical intact joint. There is only a small area of contact between the articulating surfaces. Fluid is squeezed out of this region, travelling quite long distances through the cartilage parallel to the articulating surfaces before it escapes. The rest of the cartilage is free to soak up synovial fluid, which enters as shown by relatively short paths at right angles to the surface. No part of the cartilage will get badly squeezed out, so long as the area of contact is changed from time to time. We have already seen how the cruciate ligaments ensure that the area of contact changes on both articular cartilages in the knee, when the joint is moved (page 42). Mammals do not stand absolutely still, but change their positions from time to time.

The broken lines in Fig. 28 show the results of experiments with cartilage soaked in water instead of synovial fluid. The initial coefficient of friction was higher and it rose faster than in the experiments with synovial fluid, although the cartilage was squeezed out at about the same rate. This is apparently because hyaluronic acid, a component of synovial fluid, has a boundary lubricating effect which supplements the weeping lubrication (McCutchen, 1966).

The theory of weeping lubrication at joints is simple and has striking evidence to support it, but is not universally accepted. Dintenfass (1963) stressed the complex mechanical properties of both synovial fluid and cartilage, and suggested that it might be possible to explain the lubrication of synovial joints in the light of recent investigations of engineering joints.

CHAPTER 3

ELASTICITY AND VISCOSITY

ELEMENTARY books on physics deal with the elasticity of solids such as steel and the viscosity of liquids such as water. The properties of animal materials are generally quite different, and more complicated. This is because of long, flexible molecules. The most important structural materials in animals are proteins such as collagen, polysaccharides such as chitin and inorganic salts such as calcium carbonate. Proteins and polysaccharides are high polymers: that is to say, they have very big molecules composed of large numbers of more or less similar units, joined together. Fortunately for zoologists, rubbers and plastics are also high polymers, and their properties have been studied intensively because of their commercial importance. The physics on which this chapter is based is given in much more detail than is given here, in books on rubbers and plastics such as those by Ferry (1961) and Ritchie (1965).

Elasticity

When I stretch a rubber band and hold it extended, forces act in it which tend to restore it to its original length. When I bend the blade of a knife and hold it bent, forces act which tend to straighten it again. In each case, the size of the elastic restoring forces depends on the dimensions of the band or blade, the amount it is stretched or bent, and the material of which it is made. The effect of the material is expressed by the elastic modulus of the material. Several different moduli have been defined, for different types of deformation. The most familiar is Young's modulus, which is concerned with the effects of stretching the material, or compressing it, along a single axis.

Consider the piece of material shown in Fig. 29a. Its length is *l* and we will suppose that it has uniform cross-sectional area *A*. The cross-section is of course taken at right angles to the length.

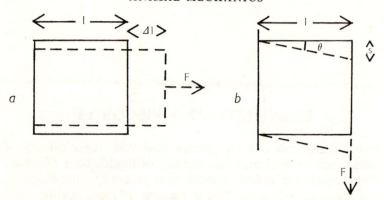

Figure 29. Diagrams illustrating (*a*) tensile strain, and (*b*) shear

One end of the piece is fixed and a force *F* is applied to the other, parallel to the length *l*. This makes the specimen longer and thinner. Its length increases to $(l + \varDelta l)$ and its cross-sectional area decreases to $(A - \varDelta A)$. If its volume remains constant as it does, more or less, for highly extensible materials like rubber

$$Al = (A - \varDelta A)(l + \varDelta l) \qquad (8)$$

The relative increase in length, $\varDelta l/l$, is called the tensile strain. Since it is a ratio of two lengths it has no dimensions. A specimen stretched to twice its initial length has a tensile strain of 1. The force per unit cross-sectional area is known as the tensile stress. There are unfortunately two different ways of defining it. It is sometimes defined as F/A and sometimes as $F/(A - \varDelta A)$. The difference between the definitions is trivial when the strain is small but important when it is large. In the case of a rubber band stretched to twice its initial length $(A - \varDelta A)$ is, by equation (8), only half as big as *A*. If force is measured in dynes and area in square centimetres, stress is given in dyn/cm². Hooke's law states that the extension $\varDelta l$ is proportional to the force *F*. This is more or less true for metals (except near the breaking point) and for small extensions of many other materials. Young's modulus is (tensile stress)/(tensile strain) and is a constant for any material of which Hooke's law is true. If the stress is in dyn/cm², Young's modulus is also in dyn/cm². If the material is being compressed along a single axis instead of being stretched, the stress and the strain are negative but their ratio is still Young's modulus. Stress is most nearly proportional to strain over large extensions, if it is taken as $F/(A - \Delta A)$, so it is best to use this definition of stress in calculations concerning large extensions.

Fig. 29b shows the same piece of material as Fig. 29a but the force is applied parallel to its end instead of at right angles to it. If the piece is sufficiently short and fat it will deform to the shape indicated by the broken line without bending appreciably. This is called shear deformation. The shear strain is the angle θ, expressed in radians. When it is small it is approximately equal to s/l. The shear stress is the force divided by the cross-sectional area. There is no ambiguity here because shearing, unlike stretching, leaves the cross-sectional area unchanged. The shear modulus is equal to (shear stress)/(shear strain).

If a piece of material is subjected to a high hydrostatic pressure it is compressed to a smaller volume. The volume strain is defined as (change in volume)/(initial volume) and the bulk modulus as (pressure)/(volume strain). It is much easier to change the shape of a piece of rubber than to change its volume: that is to say, its bulk modulus is much higher than its other moduli. For materials with this property, Young's modulus is about three times the shear modulus. For various plastics, Young's modulus lies between 2·5 and 3 times the shear modulus (Ritchie, 1965).

When an elastic material is strained work is done on it and energy is stored in it. This is a form of potential energy and can be recovered in the elastic recoil. Suppose a piece of material is stretched, and the elastic restoring force is measured at intervals during the stretch so that a graph can be drawn of force against extension (Fig. 30a). The work which is done cannot be obtained simply by multiplying a force by a distance, for the force changes as the material extends. However, if we consider only a very small part of the extension, for instance from X to Y, the force is more or less constant and the work is this force multiplied by the distance XY. In other words, the work done in stretching the material from X to Y is given by the area of the cross-hatched strip. Hence, the work done in stretching it over the

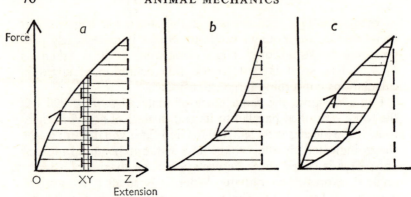

Figure 30. Diagrams illustrating mechanical hysteresis

whole distance OZ, is given by the area under the whole curve.

If the material is allowed to shorten gradually to its initial length, it does work against the force which is restraining it. If the force is measured at intervals during shortening, and plotted against the extension that remains at each stage, another graph is obtained (Fig. 30*b*) which does not quite coincide with the first one. The area under it gives the work done by the elastic restoring force.

Thus, the work done in stretching the specimen is given by the area hatched in Fig. 30*a*. The work recovered during shortening is given by the smaller area hatched in Fig. 30*b*. The difference between these areas, which is the area of the loop obtained by plotting both curves on one graph (Fig. 30*c*) gives the amount of energy lost in the process of stretching and recoil. Such loops are called hysteresis loops. The energy which is lost, is lost as heat.

The term "resilience" has been given at least three meanings, all connected with elasticity. It is used throughout this book to mean the work recovered from a material in an elastic recoil, expressed as a percentage of the work previously done in straining it. The resilience of the material which Fig. 30 describes is the area hatched in Fig. 30*b*, expressed as a percentage of the area hatched in Fig. 30*a*. The resilience of a material is not constant, but depends on the time taken to strain the specimen and allow it to recoil.

It is sometimes convenient to give the stiffness of a specimen rather than the modulus of the material it is made of. If the elastic restoring force is Sx when the specimen is strained through a distance x, the stiffness is S. Thus the stiffness in tension of the specimen shown in Fig. 29a is $F/\Delta l$ and its stiffness in shear (Fig. 29b) is F/s.

The graph shown in Fig. 30a would be a straight line if Hooke's law was followed. The force needed to extend the specimen through a distance Δl would be $S\Delta l$, where S is the stiffness of the material. The work done in stretching the specimen, as given by the area under the graph, would be $S(\Delta l)^2/2$. In more general terms, if a specimen which follows Hooke's law is deformed through a distance x, whether it be stretched, sheared, bent or twisted, and its stiffness in that kind of deformation is S, the work done is $Sx^2/2$. This expression is useful in a method of measuring resilience.

Consider a piece of clockspring whose stiffness in bending is S. It is clamped at one end. If it is bent to one side and released, it oscillates from side to side. The oscillations get gradually smaller. Suppose that in the course of these oscillations the tip of the spring swings a distance x' to one side and then a distance x'' to the other (Fig. 123b, page 280). The work needed to bend the spring to the first of these positions is $Sx'^2/2$. Some of this is recovered as kinetic energy in the recoil and serves to bend the spring to the other side. The work needed for this is $Sx''^2/2$. If air resistance can be ignored, the resilience of the spring is $100\,(x''/x')^2$. Hence, if the amplitudes of successive oscillations can be measured, the resilience can be calculated. The same value would not necessarily be obtained if the length of the spring were changed or if a weight were attached to it, so that it vibrated at a different frequency. The basic method is not, of course, limited to clocksprings. We shall see presently how it has been used to determine the resilience of an elastic protein (page 83).

The property of resilience is often measured by a different quantity, known as the loss tangent (see page 292). The resilience R is related to the loss tangent $\tan \delta$ by the equation

$$\log_e (100/R) \simeq \pi \tan \delta \qquad (9)$$

This equation is reasonably accurate when $\tan \delta < 0 \cdot 1$, but not when it is higher (Parke, 1966).

Rubbery elasticity

There is an immense difference between the elastic properties of steel and rubber. Steel breaks when it is stretched by about $0 \cdot 3\%$ of its initial length, and Young's modulus is about 2×10^{12} dyn/cm². Soft rubbers can be extended to three times the initial length and Young's modulus is around 5×10^7 dyn/cm² (Table 3). This difference in properties is due to a fundamental difference in the molecular mechanism of elasticity. Rubbery elasticity is peculiar to high polymers, and it is important to us because of the animal materials in which it occurs.

Steel consists of evenly spaced atoms. When it is deformed the elastic restoring force is almost entirely due to changes in the distances between atoms. Rubber is far from regular, and its elasticity is mainly due to another cause. It consists of long

TABLE 3

MECHANICAL PROPERTIES OF VARIOUS MATERIALS

(The values for collagen refer to stretching along the fibres, and the values for bone to stretching along the osteones)

	Young's modulus (dyn/cm²)	Authority	Tensile strength (dyn/cm²)	Authority
Metridium mesogloea	c. 3×10^4	Alexander, 1962, 1964a		
Resilin	$1 \cdot 8 \times 10^7$	Weis-Fogh, 1961a	3×10^7	Weis-Fogh, 1961b
Abductin	$1-4 \times 10^7$	Kelly and Rice, 1967		
Elastin	6×10^6	Bergel, 1961		
Collagen	10^{10}	Harkness, 1961	$5-10 \times 10^8$	Elliott, 1965
Bone	10^{11}	Smith and Walmsley, 1959	10^9	Evans, 1957
Locust cuticle	10^{11}	Jensen and Weis-Fogh, 1962	10^9	Jensen and Weis-Fogh, 1962
Lightly-vulcanized rubber	$1 \cdot 4 \times 10^7$	Ferry, 1961		
Oak	10^{11}	Hodgman, 1965	10^9	Hodgman, 1965
Mild steel	2×10^{12}	Hodgman, 1965	5×10^9	Hodgman, 1965

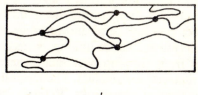

a   b

Figure 31. Diagrams illustrating the structure of a cross-linked amorphous
polymer such as rubber (*a*) unstrained, and (*b*) stretched horizontally. The
convoluted lines represent molecules

flexible molecules which are joined together here and there by
cross-links of sulphur, so that they form a three dimensional
network (Fig. 31*a*). The molecules are convoluted and heat
energy keeps them in constant Brownian motion so that they are
for ever twisting into new shapes. The distance between the ends
of any one molecule keeps changing but the length of a block
of unstressed rubber remains virtually constant because it is so
exceedingly unlikely that all the molecules will curl up or uncurl
at once. Although the length of a molecule fluctuates randomly
it has a most probable value, and at any particular instant most
of the molecules in a block of unstressed rubber will be fairly
near their most probable lengths. When the rubber is stretched,
its molecules are pulled out in the direction of stretching (Fig.
31*b*). When it is released, the random movements of the
molecules tend to bring them back to their most probable
lengths, and the rubber shortens again.

The difference between the two types of elasticity can be
described by saying that the elasticity of steel is mainly due to
changes of internal energy, and the elasticity of rubber to
changes of entropy. It is not necessary for our purposes to
know exactly what the terms "internal energy" and "entropy"
mean. They are explained in textbooks on physical chemistry
such as Glasstone and Lewis (1962). We shall use them simply as
convenient labels for the two phenomena involved in elasticity.

Elasticity which is mainly due to entropy changes is known
as rubbery elasticity. It is not a property of rubber alone but of
cross-linked high polymers in general, provided they are not too

highly crystalline and provided the temperature is not too low. The critical temperature depends on the polymer, and even rubber becomes hard and more or less inextensible at very low temperatures. Cross-linking is necessary to prevent the molecules from slipping past each other instead of uncoiling when the material is stretched. The moduli of elasticity depend on the frequency of cross-links. If there are long stretches of flexible molecule between one cross-link and the next, a lot of deformation is possible and both Young's modulus and the shear modulus are low. If the cross-links are more closely spaced less deformation is possible and the moduli are higher. It can be shown theoretically that if the density of the material is ρ and the average value of the molecular weight of the piece of molecule between one cross-link and the next is M, the shear modulus G at absolute temperature T is given by the equation

$$G = \rho RT/M \qquad (10)$$

Young's modulus is about three times this value. The quantity R in the equation is the universal gas constant, $8 \cdot 3 \times 10^7$ erg/°mol (see page 193). The equation can be used to estimate the value of M from known values of density and modulus.

Pure rubber consists of rubber molecules and nothing else. Rubbery proteins contain a large proportion of water, whose molecules lie among the protein molecules. The water contributes to the density of the material but does not affect the shear modulus. When one applies equation (10) to proteins, one should not use the density of the material, but the concentration of protein in g/cm³ of material.

This rule is probably the right one to apply when dealing with proteins which were presumably already diluted with water at the time they were cross-linked. There is a different rule for finding the molecular weight ¡between cross-links from the moduli of materials which were not diluted until after they had been cross-linked (Treloar, 1958). The process of dilution stretches out the molecules of such materials so that they are no longer randomly convoluted, and this affects the modulus. Rubber swollen with paraffin is a material of this sort.

One can with luck find out experimentally how much of the elasticity of a material is due to internal energy changes, and how much to entropy changes. The luck consists, as we shall see, in having a suitable material for the experiments. We will assume for the moment that the material is suitable.

Consider a piece of material held stretched at a constant length l while the absolute temperature T is altered. It is left long enough at each temperature to reach equilibrium, and then the elastic restoring force F is measured (this is of course the force acting on whatever is keeping the material stretched). A graph is drawn of F against T and its gradient is measured. It can be shown that the part of the elastic restoring force F which is due to entropy, is equal to T times this gradient. In mathematical terms, it is equal to $T(\partial F/\partial T)_l$ (Treloar, 1958). The rest of F must be due to internal energy. For strict accuracy, the volume of the material should be kept constant in spite of the temperature changes, by applying high pressures. This has been done in experiments with rubber (Allen, Bianchi and Price, 1963) but is quite unnecessary for most purposes, and a great deal too much trouble.

This method of distinguishing entropy elasticity from internal energy elasticity does not work for materials like table jelly whose cross-links break as the temperature rises. It does not work for materials like elastin which take up water from their surroundings or lose it as the temperature changes (though we will see on page 81 that it has been made to work for elastin by a bit of trickery).

Most materials that can stretch elastically to large strains have rubbery elasticity, due mainly to entropy changes. Not all do. Hair can be stretched to $1 \cdot 7$ times its initial length, and will spring back, but this is because the protein keratin, of which it is made, can exist in two crystalline forms. When it stretches, some of the keratin changes from a form in which its molecules run in tight helices, to one in which they are more extended (Ciferri, 1963; Feughelman, 1963). This is an entirely different phenomenon from rubbery elasticity.

Resilin

Resilin is a protein found in arthropods, serving various functions (Andersen and Weis-Fogh, 1964). Of particular interest are the structures made of resilin in the thoraxes of insects, which serve to reduce the energy which is needed for flight. Before we see how they do this, we will examine the properties which make it possible.

Resilin is hard when it is dry, but in the natural state it contains 50–60% water and is soft and rubbery. It can be made rubbery again after drying, by putting it back in water. There is an apodeme in dragonflies which is tough and inextensible for most of its length, but its middle section is almost pure resilin. It is known as the elastic tendon. Weis-Fogh (1961a) used it in some of his experiments on resilin. Though it is tiny, he was able to fasten fine nylon threads round the tough parts of the apodeme, and stretch the elastic part. The apodeme and the apparatus are shown in Fig. 32. The apodeme was fastened in a glass chamber full of water, on the stage of a microscope. The chamber was fitted with a thermocouple (b in Fig. 32c) to measure the temperature of the water and a pair of electrodes

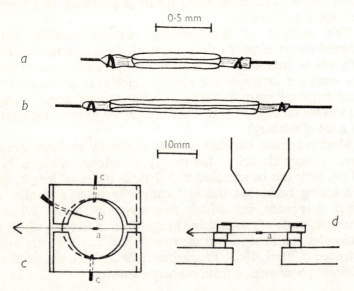

Figure 32. Illustrations of experiments on the elastic tendon of a dragonfly (*Aeshna*). (*a*) The tendon with nylon threads attached. (*b*) The same tendon, stretched. (*c*), (*d*) The observation chamber with the tendon *a* in place, seen from above and from the side. (Redrawn from Weis-Fogh, 1961a)

(c) which could be used to heat it by passing a current through it. The apodeme was stretched either by a weight attached to the nylon thread, or by an electrical device. Its length was measured by using an eyepiece micrometer on the microscope.

The resilin could be stretched to three times its initial length. Weis-Fogh stretched pieces to about twice their initial length by means of weights, and left them for days and even weeks. Their lengths remained virtually constant, but when the weight was removed each piece snapped back immediately to its initial length. Obviously the molecules of resilin must be cross-linked so that they cannot slide past each other. Young's modulus was easy to calculate, since the cross-sectional area is more or less constant along the resilin section of the apodeme. It proved to be about $1 \cdot 8 \times 10^7$ dyn/cm². We can use equation (10) (page 74) to estimate how far apart the cross-links are on the molecules. The shear modulus (G) must be about a third of Young's modulus, or 6×10^6 dyn/cm². A cubic centimetre of resilin in its natural state contains about $0 \cdot 5$ g of the dry protein, so we can take $\rho = 0 \cdot 5$ g/cm³. By putting these values in the equation, we find that the average molecular weight of the pieces of molecule between one cross-link and the next is about 2,000. This means that there must be about 23 amino acid residues between each cross-link and the next. Weis-Fogh (1961b) estimated that there were about 60 residues between cross-links, but he used the formula for swollen rubbers (see page 74). A lot of the cross-links are apparently covalent bonds joining tyrosine residues in adjacent peptide chains (Andersen, 1966).

Weis-Fogh (1961a) carried out experiments of the sort described on page 75, to find out how much of the elastic restoring force was due to entropy changes. Resilin is a very convenient subject for such experiments for it reaches equilibrium very rapidly after a temperature change and neither its water content nor its cross-links are affected by quite large changes of temperature. The results are shown in Fig. 33. The quantity λ shown on the scale at the top of the graph is the ratio of the length at which the resilin was held stretched to its initial, unstretched length. The quantity α shown at the bottom need not concern us. The graph labelled f shows the elastic restoring force at 27°C (300°K). The graph labelled $-T(\partial S/\partial \lambda)_T$ shows

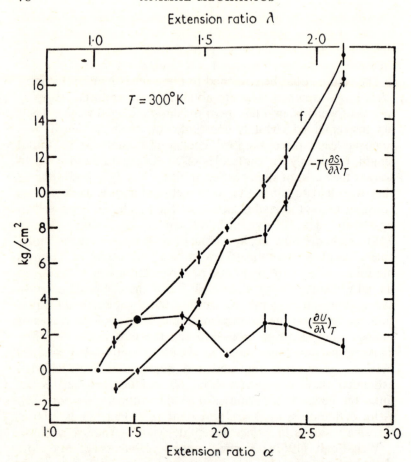

Figure 33. A graph which is explained in the text, showing the contributions of the two types of elasticity to the stress in resilin stretched by different amounts. (From Weis-Fogh, 1961b)

the part of this force that was due to entropy changes, calculated from the relationship between the force and the temperature. The third graph shows the residue of the force which must be due to internal energy changes. Most of the elastic restoring force is apparently due to entropy changes, when the resilin is stretched to 1·5 or more times its initial length. The estimates of the entropy component are probably too low because of thermal expansion. If it were possible to correct for this it would probably appear that the elasticity is mainly entropy elasticity, even at small extensions. The elasticity of resilin seems to be essentially similar to the elasticity of rubber.

Jensen and Weis-Fogh (1962) used sophisticated electrical equipment, similar to that shown in Fig. 128 (page 294), to measure the resilience of resilin. When they deformed and released the resilin 50 times a second, the resilience was 97%. When it was deformed and released at higher frequencies, the resilience was rather lower. A resilience of 97% is very high. Rubbers and plastics with similar moduli usually have resiliences of about 91%, at the most favourable rates of deformation and release (Ferry, 1961).

Thus, resilin is an elastic material with about the same modulus as ordinary soft rubber. Its elasticity depends on the same molecular mechanism as the elasticity of rubber. As rubbers go, it is exceptionally bouncy—that is, it has a very high resilience. We are now ready to look at a couple of the ways in which the properties of resilin are exploited by insects.

Insects fly by flapping their wings up and down. The moving wings have kinetic energy, and we saw on page 34 how this kinetic energy can be estimated. Between each wing stroke and the next, kinetic energy must be taken from the wings to stop them, and then kinetic energy must be supplied to them to accelerate them in the opposite direction. Resilin provides a means of using the energy that has just been taken from the wings to accelerate them again. The principle is the same as the principle of bouncing a rubber ball. A ball falling to the ground acquires kinetic energy. When it strikes the ground it is momentarily stopped and deformed. Kinetic energy is taken from it and stored as elastic potential energy. In the elastic recoil, this potential energy is converted to kinetic energy and the ball bounces upwards. It will not rise to quite the height it fell from because the resilience of the rubber is less than 100%. Nevertheless, most of the kinetic energy taken from the ball is restored to it again. The elasticity of the ball makes it possible for downward kinetic energy to be transformed to upward kinetic energy.

Insect wings have bits of resilin at their bases, arranged in such a way that the resilin is distorted when the wings are raised. If the fore-wings of a dead locust are opened, they point down. If they are pulled up the resilin is distorted, and when they are released it recoils elastically, making them swing down again. The elastic recoil of the resilin is reinforced by elastic recoil of the wall of the thorax, which is distorted as the wings are raised, and of muscles which are stretched as the wings are raised. When a locust flies, most of the kinetic energy which is taken from the wings at the top of the upstroke is stored as elastic potential energy in the resilin, thorax wall and muscles, and restored to the wings as kinetic energy for the downstroke. Much less energy has to be provided by muscular contraction, than if there were no elastic structures (Weis-Fogh, 1961c). We will return to this topic on page 264, to compare the work which must be done in providing kinetic energy with the work which must be done against aerodynamic forces, in locust flight.

Bennet-Clark and Lucey (1967) found resilin serving a different function in fleas. We saw on page 32 that when a flea jumps, the power which it needs during take off is equivalent to at least 3 W/g of muscle. This is probably well beyond the capability of any muscle, and fleas are only able to jump as high as they do because they make use of the principle of the catapult.

When a boy uses a catapult, he uses his muscles to stretch the rubber. The energy that he stores in this way, relatively slowly, is released very rapidly when he lets the rubber go. In other words, the boy exerts a relatively small power for a relatively long time and the catapult exerts a larger power for a shorter time.

Fleas have a piece of resilin at the base of the hind leg which is homologous to the resilin of the hind wing in flying insects. This resilin is so placed that it is compressed when the leg is swung up. The leg is swung up relatively slowly by a muscle and then released suddenly by a trip mechanism. The resilin recoils rapidly, releasing far more power than the muscles could, just as a catapult can release more power than its user's muscles. The recoil swings the leg down very fast so that it kicks the flea into the air.

Brown (1967) has shown that the locust's jump also depends on the principle of the catapult, but the elastic energy is stored in the main jumping muscle, not in resilin. The elastic properties of muscle are described on page 107.

Elastin

Elastin is an elastic protein found in vertebrates. It is present as thin strands in areolar connective tissue. It forms quite a large proportion of the material in the walls of arteries, especially near the heart (Harkness, Harkness and McDonald, 1957). The ligamentum nuchae which runs along the top of the neck in ungulates (Fig. 34) is almost pure elastin.

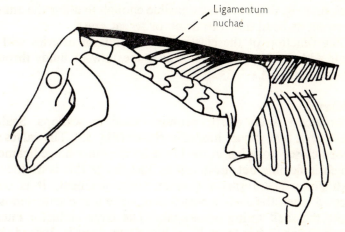

Figure 34. Part of a horse skeleton, showing the position of the ligamentum nuchae

Elastin can be stretched a long way, like rubber, and snaps back to its initial length when it is released. Young's modulus is about 6×10^6 dyn/cm² (Bergel, 1961), or a little lower than for resilin (Table 3, page 72). It is hard to show how much of the elasticity is due to entropy, because elastin loses water and shrinks when it is heated. However, Hoeve and Flory (1958) carried out experiments on elastin in a mixture of water and glycol in which it neither swells nor shrinks on heating. They claimed to have shown that its elasticity is almost entirely due to entropy and they were probably right, though the validity of their technique has been questioned (see Hoeve and Flory,

1962). Elastin is very like resilin in its properties but quite different in chemical composition (Serafini-Fracassini and Tristram, 1966). The similarities of the materials are due solely to their both having long, flexible, cross-linked molecules, like rubber.

The function of the ligamentum nuchae of ungulates can also be described here. It is particularly large in horses and cattle, which have large, heavy heads. If they depended entirely on muscles to hold their heads up, a good deal of energy would be used, maintaining the necessary tension in the muscles. The ligamentum nuchae is arranged so as to help hold the head and neck up (Fig. 34), but it is extensible enough to allow the animal to swing its head down to drink or to eat grass.

The function of the elastin in the walls of arteries will be dealt with later, when we consider the flow of fluids through tubes (page 211).

Abductin

Abductin is an elastic protein found in scallops. Scallop shells have two valves, hinged together (Fig. 35). The valves are joined closely together at the hinge by the outer hinge ligament, which is flexible but inextensible. Just inside this is a block of abductin which forms the inner hinge ligament. It is compressed when the shell is closed and, since it is elastic, tends to make the shell spring open again. The large adductor muscle closes the shell but there is no muscle to open it. Indeed, it is hard to see how a shell could be opened by a muscle inside it.

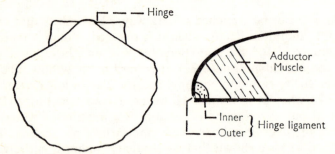

Figure 35. Lateral view and diagrammatic transverse section of a scallop (*Pecten*), showing the hinge and the adductor muscle

The inner hinge ligament serves as the antagonist of the adductor, opening the shell when the adductor relaxes.

The inner hinge ligament bounces like rubber when it is removed from the shell and dropped. Young's modulus can be determined most conveniently with it still in place in the shell. The adductor can be removed and it is then easy to measure the force needed to close the valves to any particular position. Estimates of Young's modulus, from measurements on different species of scallop, range from $1 \cdot 3 \times 10^7$ to 4×10^7 dyn/cm² (Trueman, 1953; Alexander, 1966c; Kelly and Rice, 1967). The cross-links must be spaced about as far apart as in resilin, or rather closer. It has been shown that the elasticity is true rubbery elasticity, due mainly to entropy changes, by experiments of the kind described on page 75 (Alexander, 1966c).

Scallops are able to swim, by opening and closing their valves very rapidly. I have watched the small scallop *Chlamys* swimming and estimate that it opens and closes its valves about three times a second. The greater the resilience of the abductin when it is compressed and released with this frequency, the less energy will be needed for swimming. Fig. 36 shows how the resilience was measured (Alexander, 1966c; the method had previously been used by Jensen and Weis-Fogh, 1962, to

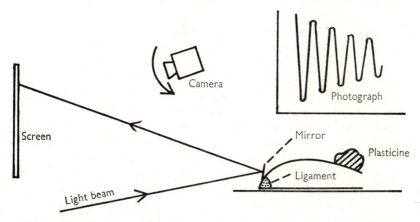

Figure 36. The apparatus used to measure the resilience of the inner hinge ligament of *Chlamys* and (inset) a diagram of one of the photographic records

measure the resilience of insect cuticle). The body was removed from the shell of a specimen of *Chlamys*, leaving the hinge intact. One valve was clamped to a bench and a small mirror was attached to the other. The laboratory was darkened and a narrow beam of light was projected on to the mirror, so that it was reflected on to a screen. When the free valve of the shell was pressed down and released, it oscillated up and down with gradually diminishing amplitude. By fixing plasticine to it the frequency of the oscillations could be adjusted to match the frequency of the swimming movements. As the shell oscillated, the light spot moved up and down on the screen. A camera on a tripod was trained on the screen. Its shutter was opened before the shell was set oscillating, and kept open as the oscillations decayed. As they decayed, the camera was turned slowly, so that the light spot traced an undulating line on the photographic plate, as shown at the right of Fig. 36. The amplitudes of successive oscillations were measured from the plate, and the resilience calculated in the manner described on page 71. It proved to be about 91 %. This is in the usual range of values for rubbery materials, but lower than the values for resilin.

Abductin is very like resilin and elastin in its physical properties, but quite different from either in chemical composition (Kelly and Rice, 1967; Andersen, 1967).

Hysteresis of bivalve hinge ligaments

The scallops are exceptional bivalves. They swim by opening and closing their shells repeatedly and very rapidly, **and** have an inner hinge ligament with a high resilience. Other bivalve molluscs do not make fast, repeated movements, and it is perhaps of less consequence to them how much of the work done by their adductor muscles is lost as heat. Trueman (1953) investigated the properties of the hinge ligaments of various bivalves, by opening and closing their shells and obtaining hysteresis loops (see page 69). He obtained much more slender loops with scallops than with other bivalves. The resilience estimated from his curves is 90 % or more for scallops, but 80 % or less for other bivalves. Each cycle of closing and opening the shell took several minutes, so these figures refer to very much slower movements than the swimming movements of scallops.

Viscosity

Fig. 37a is very like Fig. 29b but it shows a shearing stress being applied to a viscous fluid instead of to an elastic solid. A layer of fluid represented by stippling is sandwiched between two flat plates. The thickness of the layer is l and the area of the plates is A. The lower plate is fixed and a horizontal force F is applied to the upper one. The shearing stress is F/A. The upper plate moves horizontally, and its motion is resisted by the viscosity of the fluid. The layer of fluid immediately in contact with the lower plate remains stationary but the layer in contact with the upper plate moves with the upper plate. At intermediate levels the velocity of the fluid is proportional to the distance from the lower plate. If the stress remains constant the upper plate moves with constant velocity v and there is a

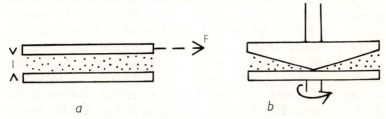

Figure 37. (*a*) A diagram illustrating the definition of viscosity in the text. (*b*) A schematic section through a cone and plate viscometer

velocity gradient v/l in the liquid. For many liquids, the velocity gradient is proportional to the stress. The more viscous the fluid, the smaller will the velocity gradient be, and the viscosity of the fluid is defined as (shearing stress)/(velocity gradient). The velocity gradient can be thought of as the rate at which the liquid is being strained (for a definition of shearing strain, see page 69), and the definition of viscosity can be re-worded as (shearing stress)/(rate of strain). If v is expressed in cm/s, l in cm, F in dyn and A in cm², the viscosity is obtained in dyn s/cm² or poises. The viscosity of water at 20°C is 0·01 poise and that of glycerin is about 15 poise.

There are various ways of measuring the viscosity of liquids. In one method, the liquid is forced through a capillary tube by a known pressure and the rate of flow is measured. The viscosity can be calculated by means of equation (26) (page 210). Other methods involve a layer of fluid sandwiched between a stationary disc and a rotating one. The apparatus shown in Fig. 37b is called a cone and plate viscometer. The plate below the liquid is rotated at a constant speed and the viscosity of the liquid tends to make the cone rotate as well. The couple required to prevent it from rotating is measured and the viscosity can be calculated (see Ritchie, 1965). The advantage of having the liquid sandwiched between a cone and a plate instead of between two flat plates, is that the velocity gradient is constant throughout the liquid. The velocity of the rotating plate increases with distance from its shaft, but the thickness of the liquid increases proportionately.

In liquids like water and glycerin the velocity gradient is proportional to the shearing stress so that the viscosity at any given temperature is constant. They are called Newtonian liquids. In high polymers and solutions of high polymers, the velocity gradient is not proportional to the shearing stress, but the viscosity generally decreases as the stress and the velocity gradient increase (Lodge, 1964). This is because the molecules are long. If a long molecule was in a vertical position in the fluid represented in Fig. 37a, its upper end would be carried along faster than its lower end by the surrounding fluid, and it would tend to rotate to a horizontal position. In general, long molecules in a velocity gradient tend to lie in planes at right angles to the gradient so that the whole of the molecule lies in a layer of liquid travelling with the same velocity. Random movements of the molecules disturb the arrangement but the greater the velocity gradient, the better the arrangement and the less one layer of liquid interferes with the movement of the next. Hence, the viscosity tends to decrease as the velocity gradient increases.

Synovial fluid

Synovial fluid fills the cavities of the synovial joints of mammals (page 62). It is similar in constitution to blood plasma, but it contains less protein and some hyaluronic acid. This is a polysaccharide with long molecules apparently combined with protein. Synovial fluid is much more viscous than blood and this seems to be due to the hyaluronic acid. An enzyme which destroys the hyaluronic acid, greatly reduces the viscosity (Ogston and Stanier, 1953b).

Various investigators have found that the viscosity of synovial fluid falls as the velocity gradient rises. For instance, Davies (1966) studied fluid from cattle, using a cone and plate viscometer. Fluid from the joint between the radius and the carpals had a viscosity of about 50 poise in a velocity gradient of 0·1/s and about 0·1 poise in a velocity gradient of 1,000/s. As the velocity gradient increased the viscosity fell from 5,000 times the viscosity of water to only 10 times the viscosity of water. Fluid from another joint which contained less hyaluronic acid had lower viscosities.

Synovial fluid has another striking property. Ogston and Stainier (1953a) demonstrated it in a simple experiment. A drop of synovial fluid, or of another liquid, was placed on an optically flat surface and a convex lens was set on top. The lens was pressed down and the interference patterns known as Newton's rings were used to measure the distance between the lens and the optical flat (for an explanation of Newton's rings see Nelkon and Parker, 1965). When the experiment was done with water, the lens was easily pressed into contact with the optical flat. The same happened with blood plasma diluted to have the same protein composition as synovial fluid. With glycerol, the lens moved very slowly down towards the flat. With synovial fluid, the lens stopped moving while it was still some distance above the optical flat. The distance depended on the load pressing it down. Moreover, when the load was removed, the lens moved slightly up again. This was apparently due to elastic recoil. Synovial fluid seems to have elastic properties, as well as viscous ones. The very thin layer of fluid that remains between the lens and the optical flat, behaves very much as if it were a thin layer of rubber.

There is an even simpler way of showing that synovial fluid is elastic as well as viscous. If one swirls it in a flask and stops the movement suddenly, the elastic recoil sets it rotating for a moment in the opposite direction.

There is nothing very mysterious about the properties of synovial fluid, though the idea of an elastic liquid may be hard to grasp. Many other dilute solutions of high polymers have viscosities which decrease as the velocity gradient increases, and are elastic (Lodge, 1964). The next section but one of this chapter is about materials which are at once viscous and elastic.

The properties of synovial fluid may make it impossible to squeeze it out entirely from between the articulating surfaces in joints, just as it cannot be squeezed from between a lens and an optical flat. This would obviously be a valuable property for a lubricant.

Thixotropy

Certain kinds of paint become more fluid when they are stirred and take some time to revert to their original state when stirring stops. This property is called thixotropy. It is not quite the same thing as non-Newtonian viscosity, though elastic liquids often have both properties (Lodge, 1964). The viscosity of a non-Newtonian liquid decreases as the velocity gradient increases. The viscosity of a thixotropic liquid decreases gradually in a constant velocity gradient, and remains low for some time after shearing ends. Synovial fluid seems to be thixotropic as well as having non-Newtonian viscosity (Davies, 1966).

Wet sand has a property which resembles thixotropy. A mixture of sand and water which contains more than about 65% by volume of sand is a stiff paste. A mixture which contains less than about 60% behaves as a Newtonian liquid (Cottrell, 1964). If wet sand is poked and stirred, the disturbed sand tends to take up water from the surrounding sand and to become more fluid. This property may be exploited by the lugworm *Arenicola*, which burrows in wet sand (Chapman and Newell, 1947). It probes and scrapes away the sand in front of it by repeated movements of its proboscis, which probably make the sand take up water and become temporarily more fluid. The pressure in the surrounding sand falls as water is withdrawn from it (Trueman, 1966c). The disturbed sand must quickly become stiff again, leaving the burrow with firm walls.

Visco-elastic properties

When an elastic material of shear modulus G is sheared to a strain θ, the stress is $G\theta$ (page 69). When a liquid of viscosity η

is sheared so that the rate of strain (or velocity gradient) is $d\theta/dt$ the stress is $\eta \, d\theta/dt$ (page 85). $d\theta/dt$, is a mathematical expression meaning the gradient of a graph of θ against time, t). Imagine a material which is both elastic and viscous, so that shear is resisted by both an elastic restoring force and a viscous resistance. The total stress F/A is the sum of a component due to elasticity and a component due to viscosity, so that

$$F/A = G\theta + \eta \, d\theta/dt \tag{11}$$

When stress is first applied, the strain θ will be zero and the rate of strain $d\theta/dt$ will be relatively high. As time goes on and θ increases, $G\theta$ will make up an ever increasing proportion of the stress, if the stress remains constant, so that $d\theta/dt$ must decrease. The strain will increase at a gradually diminishing rate, gradually approaching but never quite reaching the equilibrium value at which $G\theta$ would be equal to the stress. Equation (11) can be solved by calculus, giving the equation

$$\theta = (1 - e^{-Gt/\eta}) \, F/AG$$

The ratio of the viscosity to the modulus, η/G, has the dimensions of time and is known as the retardation time τ. The equation can be re-written

$$\theta = (1 - e^{-t/\tau}) \, F/AG \tag{12}$$

This is an equation for shearing but it could be converted to an equation for stretching by substituting Young's modulus for the shear modulus (G) and tensile strain for shear strain (θ). Stretching is opposed by viscosity as well as elastic restoring forces, because it involves shearing (page 139).

Real visco-elastic materials are seldom as simple as the one we have been imagining. When a constant stress is applied to a polymer, three components of strain can usually be distinguished. They are shown separately in Fig. 38a to c. Each of the graphs in this figure is a graph of strain against time and in each it is supposed that a constant stress is applied in the time interval indicated by the broken lines. The components are

(a) Instantaneous elastic strain, shown in Fig. 38a: this strain occurs immediately the stress is applied and disappears as soon as it is removed.

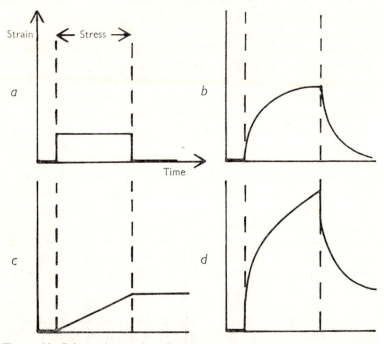

Figure 38. Schematic graphs of strain against time, for a visco-elastic material which was subject to a constant stress during the interval of time marked by the broken lines on each graph, but no stress before or afterwards. (*a*) Instantaneous elastic strain, (*b*) slow elastic strain, (*c*) viscous flow, and (*d*) total strain

(*b*) Slow elastic strain, shown in Fig. 38*b*: this increases at a gradually diminishing rate, so long as the strain is applied. This is the sort of strain represented by equation (12) but it does not normally occur in strict accordance with that simple equation. The difference between the slow elastic behaviour of a real material and the behaviour predicted by the equation, can be expressed by saying that the real material does not have a single retardation time, but a whole spectrum of retardation times (Ritchie, 1965). When the stress is removed the slow elastic strain disappears, at a gradually diminishing rate.

(*c*) Viscous flow, shown in Fig. 38*c*: this component of the strain increases at a constant rate while the stress is applied, and remains after the stress has been removed. This component is absent when the polymer is cross-linked, unless the cross-links are weak enough to be broken by the stress.

Fig. 38d shows the three components of strain added together. This is the sort of graph that is obtained when a polymeric material is stressed and then released. The exact shape of the graph depends on the material. For instance, resilin has strong cross-links and does not flow, and the retardation times of its slow elastic component are so short that in ordinary experiments the whole of its elasticity seems instantaneous.

When viscous flow occurs, elastic recovery cannot be complete. If a material is strained for a very short time and then released, relatively little flow will occur and recovery may be virtually complete, even though the material is not cross-linked. If it is strained for a long time, a lot of flow will occur and recovery will be very incomplete. In other words, a material may behave like a rubbery solid when it is stressed briefly, and like a viscous liquid when stress is prolonged. This combination of properties is shown by "bouncing putty" (polydimethyl-siloxane, a high polymer). If a ball of bouncing putty is dropped on the floor it bounces like rubber, but if it is set on a table and left for a while, it flows very gradually into a pancake shape.

Mesogloea

The sea anemone *Metridium* can change its size enormously, by inflating itself with water or contracting and driving the water out. We will see how this is done in Chapter 5 (page 168). The size changes are made possible by the properties of the very extensible body wall, which we will consider now. The body wall consists of two layers of cells with a gelatinous mesogloea between them. The mechanical properties of the body wall probably depend mainly on the mesogloea.

Fig. 39a shows apparatus used to investigate these mechanical properties (Alexander, 1962). A strip of body wall was cut from an anemone which had been narcotized with magnesium chloride to put its muscles out of action. It was attached to the short end of the lever shown in the figure, and stretched by a weight on the other side of the pivot. The weight sinks into water as the specimen stretches, and because of the upthrust of

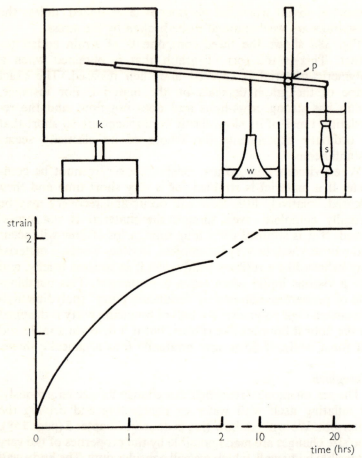

Figure 39. (*a*) The apparatus used to stretch the body walls of sea anemones
at constant stress
k, kymograph drum; *p*, pivot; *s*, specimen; *w*, weight
(*b*) A graph of strain against time, from an experiment with *Metridium*

the water on it, exerts less and less force on the specimen as it
sinks. Its shape is so designed that the force decreases while the
specimen stretches and gets thinner, in such a way as to keep the
tensile stress ($F/(A - \Delta A)$, see page 68) constant. A glass
stylus on the end of the lever records the stretching of the
specimen, by marking a line on a rotating kymograph drum.

Fig. 39b shows the result of a typical experiment. When the stress was applied at zero time on the graph there was a small instantaneous strain, followed by a much larger slow elastic strain. The specimen stretched to a strain of 2 (i.e. to three times its initial length) and was still stretching, though very slowly, after 23 h. There was little or no viscous flow, so the graph is not exactly like Fig. 38d. Elastic recovery was followed in some experiments, and was found to be slow, but more or less complete.

The very slow stretching shows that the viscosity of the mesogloea is high, though its modulus of elasticity is low. This means that it is very much harder to stretch the mesogloea quickly, than to stretch it slowly. This seems to have an advantage for the animal. It means that the body wall can withstand sudden knocks reasonably well, but does not need much pressure to inflate it when the animal enlarges slowly.

The mesogloea of sea anemones is an extremely unusual material in that its slow elastic behaviour is quite accurately described by equation (12). It is not clear why this should be, but the mesogloea of jellyfish is much more normal, with a wide spectrum of retardation times (Alexander, 1964a). Both have very low elastic moduli as one might expect, for they are dilute jellies containing a very high proportion of water (Table 3, page 72).

Fibres

Fibres form a distinctive class of polymeric materials. Animal fibres such as collagen, plant fibres and synthetic textile fibres all belong to this class. Elastin "fibres" do not: they are merely thin strands of a rubbery material. It can be shown by X-ray diffraction that fibres contain both crystalline regions where the molecules are arranged in orderly patterns, and amorphous regions where they are arranged randomly. It can be shown by electron microscopy that they contain very fine, long fibrils which are typically about 100 Å wide (Hearle and Peters, 1963. 1 Å $= 10^{-8}$ cm). The fibrils usually run more or less parallel to the long axis of the fibre but follow helical paths in the walls of the long cells that form plant fibres.

Hearle (1958, 1963a) has suggested that many fibres may have

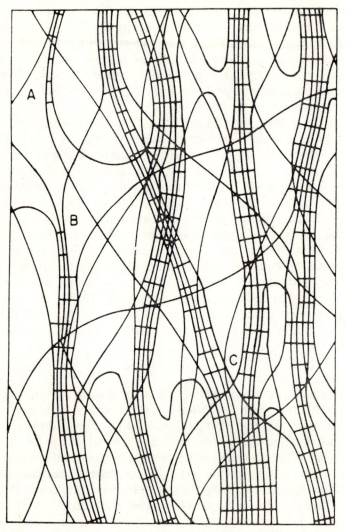

Figure 40. A diagram illustrating the fringed fibril theory of fibre structure.
(From Hearle, 1958)

their molecules arranged in the manner shown in Fig. 40. This is described as a fringed fibril arrangement. It includes regions where bundles of molecules run parallel to each other, bound together in a crystalline structure. These are the fibrils. It also includes amorphous regions between the fibrils, where individual molecules run in all directions at random. Successive sections of one long molecule may run in a fibril, separate from it and wander through the amorphous matrix, join another fibril, and so on. Though each molecule only runs for a short distance in any one fibril, the fibrils themselves are very long. It is supposed that fibril ends such as A and B (Fig. 40) must occur here and there in a fibre, and that fibrils may sometimes branch as at C. It is assumed that it must be possible to distort the crystal lattice so that the fibrils can curve gently, as the fibrils seen with the electron microscope do. It seems likely that many fibres have this sort of structure but rayon and some other fibrous materials are believed to have their molecules arranged differently.

Crystals and presumably fibrils are much less extensible than rubbery materials. Stretching them involves altering the distances between atoms rather than straightening out flexible molecules. Young's modulus for fibres stretched lengthwise is between about 10^9 and 10^{11} dyn/cm^2, which is much more than the values of 10^7–10^8 dyn/cm^2 which are usual for rubbery materials. Nevertheless, the moduli of fibres are much lower than the moduli of crystals. Hearle (1936b) supposes that when a fibre is stretched, its fibrils get straightened out as well as stretched. Straightening would distort the amorphous material between the fibrils, and would be resisted by the viscous and elastic properties of the matrix.

A material whose molecules run at random has the same Young's modulus, whatever the direction of stretching, and the same shear modulus, whatever the direction of shearing. This is not true of fibres, whose molecules are more or less aligned. For instance, Pinnock and Ward (1966) determined Young's modulus for Terylene fibres by stretching them lengthwise, and by squashing them transversely. Undrawn Terylene, which has randomly directed molecules, has equal longitudinal and transverse moduli. Drawn Terylene fibres, which have their molecules more or less aligned along the fibre, have greater longitudinal than transverse moduli. The greater the degree of alignment, the greater the difference between the moduli. The effects of alignment are not quite the same in all materials but always seem to involve an increase in Young's modulus for stretching in the direction of alignment (Ward and Pinnock, 1966; Pinnock and Ward, 1966). This is not surprising. Molecules which are already partly aligned in the direction of stretching, cannot be straightened out as much as those which are arranged randomly.

Collagen

The collagens are a group of fibrous proteins which are particularly important in vertebrates. For instance, about 20% of the protein in a mouse is collagen (Harkness, 1961). Collagens are the main constituents of tendon, of most ligaments and of the dermis, and major constituents of bone and cartilage. They are present in many other tissues, and in many invertebrates. The mesogloea of coelenterates (page 91) consists largely of a collagen. The tough cuticle of nematode worms contains a collagen. Collagens from different animals and even from different organs in the same animal are not identical. They are recognized as collagens by general similarity of composition and by some diagnostic features of their X-ray diffraction patterns.

Most collagens have fibrils whose structure repeats regularly every 600–700 Å along their length. This can be demonstrated by X-ray diffraction, and can be seen in electron micrographs as a pattern of alternate light and dark bands across the fibril. Collagen fibres can be stretched by 10–20% of their length before they break, and it can be shown by making X-ray diffraction patterns with stretched fibres that 10% extension of a fibre increases the spacing of the pattern by 9% (Cowan, North and Randall, 1955). This means that the fibrils are actually stretched by 9% and only a small part of the extension is due to straightening of the fibrils. It would be most unlikely that the fibrils could stretch by as much as 9% if they were continuous strands of crystalline material, like the fibrils shown in Fig. 40. However, there are other reasons for believing that the 600–700 Å pattern involves alternation along the fibril of highly crystalline regions and regions formed from different amino acids in which the molecules run more nearly at random. The stretching must occur mainly in these latter regions. Young's modulus is about 10^{10} dyn/cm², for collagen fibres from human skin (Harkness, 1961). This is much higher than the moduli of rubbery materials, but lies in the usual range for fibres. The high modulus of collagen makes it a very suitable material for tendons and ligaments which must withstand large forces without stretching too much.

When collagen fibres are heated they shrink to about one-third of their initial length. This happens to mammal collagens at about 65°C. It is the basis of the technique of making shrunken human heads (Harkness, 1966): the skull is removed, and the collagen of the dermis is made to shrink by heat. The shrinkage of collagen is due to breakdown of the crystalline structure. The fibrils dissociate and the molecules cease to run parallel but to run at random in all directions. Shrunken collagen gives no X-ray diffraction pattern, since it is no longer crystalline. It can be stretched to about its original length and recoils elastically. Young's modulus is about 10^7 dyn/cm². Heat has changed the fibrous material to a rubbery one. The change is thermodynamically similar to the melting of a solid (Oth, Dumitru, Spurr and Flory, 1957).

When a mammal gives birth, various tissues that contain collagen have to stretch enormously. One of them is the wall of the uterine cervix, where the two horns of the uterus enter the vagina. In rats, it contains 5–10% of collagen by weight. The Harknesses have investigated its physical properties (Harkness and Harkness, 1959a). They used cervixes cut from rats at various stages of pregnancy. They slipped a rod through each of the canals leading to the two horns of the uterus, fixed one rod and applied a force to the other. They recorded the extension of the cervix, stretched by a constant load (Fig. 41). Cervixes from rats which were not pregnant, or had been pregnant for up to 12 days, were relatively inextensible. They stretched only a little, at a gradually decreasing rate, and reached a constant length within half an hour. The amount they were stretched by any particular load was about the

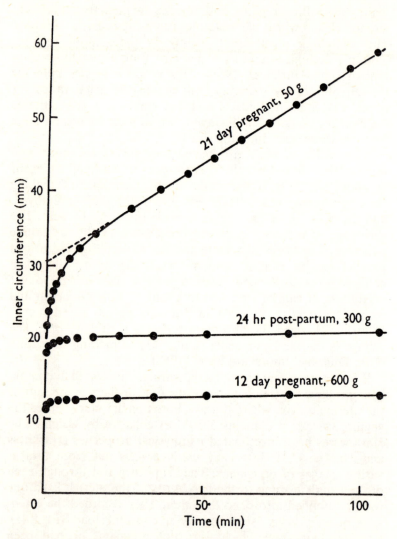

Figure 41. Graphs showing the extension of the cervixes of rats in various stages of pregnancy, stretched by the loads indicated. (From Harkness and Harkness, 1959)

amount one would expect if individual collagen fibres formed loops running right round the cervix or if they were joined firmly together to form a continuous network running round it. Later in pregnancy the cervix enlarges and becomes much more extensible. It no longer reached a constant length when it was stretched by a constant load, but the rate of stretching settled down to a constant value. It behaved as though the collagen fibres did not form a continuous network but were free to slide past each other and allow the cervix to stretch indefinitely. It seems that stretching is not resisted by an elastic restoring force in the fibres, so much as by the viscosity of some material between the fibres which is sheared as they slide past each other. There is some other evidence to support this idea. If the cervix is treated with the enzyme trypsin it stretches much faster, although trypsin does not attack collagen (Harkness and Harkness, 1959b). Within a day after the young have been born the cervix reverts to its original properties, though it is still larger than it was before pregnancy. **139478**

The main organic constituent of the mesogloea of sea anemones is a collagen, which seems to be present as fibres running in all directions through the mesogloea (Chapman, 1966). The mesogloea is elastic and highly extensible (page 91). If the collagen fibres are responsible for the elasticity and do not merely slide past each other when the mesogloea stretches, they must have quite different properties from typical vertebrate collagen fibres.

Crystallization due to strain

Raw (unvulcanized) rubber is a high polymer which is not cross-linked. If a piece is stretched to several times its initial length and held extended, the stress gradually dies away. When it is eventually released it does not recoil to its original length. This is not due to viscous flow, for it can be made to recoil by heating it. It is due to crystallization. Stretching extends the molecules so that they tend to run parallel to each other, and groups of parallel molecules tend to form crystalline regions in the rubber. The rubber cannot recoil until the crystals break up, and they only do this on heating (Treloar, 1958).

Similar crystallization can occur when liquid polymers are sheared, for instance by forcing them through tubes. The velocity of a liquid in a tube is highest at the centre and low near the walls (page 211). If a long molecule enters a tube sideways on, the end of the molecule which is nearer the wall of the tube will tend to lag behind the end that is nearer the centre, so that the molecule will come to point in the direction of flow. Long molecules flowing along a tube thus get aligned, more or less parallel to each other, and crystallization tends to occur. The material that enters the tube as a liquid may leave it as a fibre, with its molecules cross-linked and held aligned by being joined together in crystalline fibrils.

Silk

Silk is produced by various insect larvae and a few adult insects (Wigglesworth, 1965). Commercial silk comes from the silkworm, which is the larva of a moth, *Bombyx*. The silk forms a protective cocoon in which the silkworm pupates. It is produced by a pair of glands which have a common opening through the slender nozzle of the spinneret, near the silkworm's mouth.

The silk contains two proteins. The commercially important one is fibroin. The other is sericin, a gummy material which dissolves in warm water and is removed in the manufacture of silk thread. The strand of silk formed by the silkworm consists of two tough fibres of fibroin (one from each gland) enveloped in sericin.

Fibroin taken directly from the silk glands is soluble in water. X-ray diffraction patterns show that it is non-crystalline. When it passes through the fine nozzle of the spinneret its molecules become aligned and crystallize to form the fibres. It is drawn through the spinneret by tension in the thread which has already been formed. The thread adheres to surrounding objects and to the rest of the cocoon because of the stickiness of the sericin, and the silkworm moves its head to pull more silk through the spinneret. About 1 cm of silk is formed per second. Solutions of fibroin can also be made to crystallize artificially, for instance, by shearing in a cone and plate viscometer. If the velocity gradient in the viscometer is high enough, insoluble fibrous clots are formed (Iizuka, 1966).

The properties of silk are those of a fairly typical fibre. Young's modulus is about 10^{11} dyn/cm^2 and it breaks when it is stretched by about 20%. The material that spiders' webs are made of is very similar to silk, both in the way the fibres are formed and in the properties of the fibres (Savory, 1952).

Fabrics

A woven fabric consists of two sets of fibres which cross each other at right angles. An important property of fabrics is easily demonstrated with a handkerchief. It cannot be stretched much by pulling parallel to an edge, but it can be stretched quite a lot by pulling along a diagonal. The fibres are more or less inextensible but the angle between the two sets of fibres can change, and this makes diagonal stretching possible.

Shape changes of nemerteans and flatworms

Amphiporus is a nemertean worm which lives on British shores. It can extend its body so that it is long and thin, or contract it so that it is short and fat. The maximum length of a specimen is about five times its minimum length. Cowey (1952) investigated the anatomy of *Amphiporus* to find out why it can change its shape so much, and what limits the changes.

Amphiporus has a thick basement membrane which invests its whole body immediately under the epithelium. It can be shown by impregnating microscope sections with silver that this basement membrane consists of a series of layers of fine fibres. In each layer the fibres all run parallel to each other. All the layers have their fibres at the same angle to the long axis of the body but successive layers have them inclined at this angle to the left and to the right. The fibres may be collagen. Fibres in fresh material can be broken by pulling with a micromanipulator, but they cannot be stretched appreciably. In sections of *Amphiporus* which were extended when they were fixed, the fibres make a small angle with the long axis of the body. In sections of *Amphiporus* which were contracted, they make a large one. Extension and contraction of the worm distorts the basement membrane in just the same way as a diagonal pull distorts a handkerchief.

Cowey (1952) considered how the basement membrane

would limit the worm's ability to change its shape, if the fibres were utterly inextensible. The fibres run in helices round the body (I am using the term "helix" loosely: the paths would be true helices if the body was circular in cross-section, but it is usually more or less flattened). Consider the basement membrane of a section of the body just long enough for each helical fibre to make a complete turn round the body (Fig. 42a). Let this section have length l when the fibres make an angle θ with the long axis of the body. l and θ are variables but the length of fibre, D, which makes a complete turn round the body is constant. If one imagines the basement membrane of Fig. 42a cut along its length and unrolled (Fig. 42b) it is apparent that

$$l = D \cos \theta$$

and the circumference of the body is $D \sin \theta$.

The volume of body which our section of basement membrane can contain, will be greatest when it is cylindrical and less when it is flattened. When it is cylindrical its radius will be r where

$$2\pi r = D \sin \theta$$
$$r = D \sin \theta / 2\pi$$

The corresponding volume v is given by

$$v = \pi r^2 l$$
$$= \pi (D \sin \theta / 2\pi)^2 \, D \cos \theta$$
$$= D^3 \sin^2 \theta \cos \theta / 4\pi$$

Fig. 42c shows how v varies with θ. So as to avoid the need to assume a particular value of D, values of v itself have not been plotted, but of $4\pi v / D^3$ ($= \sin^2 \theta \cos \theta$). v is of course zero when $\theta = 0$ (which makes $r = 0$) and when $\theta = 90°$ (which makes $l = 0$). It has a maximum when $\theta = 55°$.

Not all values of θ are possible. The basement membrane must be able to contain the worm's body. Suppose that the part of the body contained in our section of basement membrane has volume V. The only possible shapes will be those for which $v \geqslant V$. The worm will have a maximum length at which $v = V$ and the basement membrane is only just able to contain the body with its cross-section circular. It will have a minimum

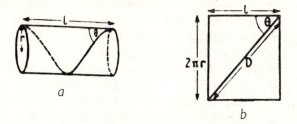

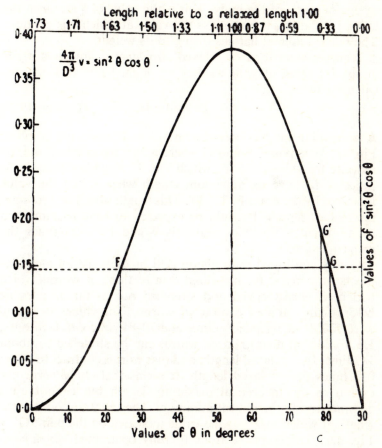

Figure 42. Diagrams illustrating the discussion in the text of the ability of nemertean and turbellarian worms to change their shape. (From Clark and Cowey, 1958)

length at which, again, $v = V$ and the cross-section must be circular to contain the worm. All intermediate lengths will be possible. At these lengths, $v > V$ and the worm will not have a circular cross-section, but a flattened one.

If a specimen of *Amphiporus* is anaesthetized, it attains a certain length. If it is then stretched and released, it quickly returns to this relaxed length. If it is fixed at this length and sectioned, it is found that the fibres are running at about 55° to the long axis of the body. This is the value of θ at which the animal is flattest, because it is the value at which v is greatest. The relaxed animal flattens owing to its weight.

Transverse sections of relaxed *Amphiporus* are on average about 5·1 times as wide as they are high. From this one can calculate that

$$4\pi V/D^3 = 0 \cdot 144 \qquad \text{(Cowey, 1952)}$$

A horizontal line has been drawn at this level on the graph in Fig. 42c. The points F and G where it cuts the graph of $4\pi v/D^3$ indicate the values of θ for which $v = V$. *Amphiporus* could be expected to have its maximum length when θ had the value indicated by the point F, 23° 30'. This length would be 1·6 times the relaxed length. It could be expected to have its minimum length when θ was 82°, and this would be 0·25 times the relaxed length.

Cowey confirmed these theoretical predictions, by measurements on worms. He measured θ in sections of worms which had been anaesthetized and stretched out as far as possible before fixing, and of contracted worms. The values he found were 22–24° in extended worms and 70–81° in contracted ones. He found that anaesthetized worms can be stretched to about 1·7 times their relaxed length, and that worms contract to about 0·3 times their relaxed length. It seems that *Amphiporus* can extend to the theoretical maximum length but cannot quite contract to the theoretical minimum. They contract only to the length at which v/D^3 has the value indicated by the point G' in Fig. 42c. Transverse sections of fully contracted *Amphiporus* are not quite circular.

Other nemerteans, and turbellarian flatworms, have basement membranes like *Amphiporus*. Some of them were studied by Clark and Cowey (1958). Two species had higher values of V/D^3 than *Amphiporus* and were therefore less flattened at the relaxed length and could not extend or contract so much. Their actual changes of length agreed well with the calculated ones. Some other species including triclad flatworms had much lower values of V/D^3 than *Amphiporus* and were therefore much flatter. If their changes in length had been limited only by the basement membrane, they would have been capable of even greater changes of length. In fact, they could not change their lengths as much. They have some connective tissue fibres among their muscles, and these may limit their changes of length.

Striated muscle

Resting muscle is a visco-elastic material with quite ordinary properties (Buchthal and Kaiser, 1951; Rack, 1966). The really interesting property of muscle is that of contraction.

Fig. 43a shows apparatus which has been used to investigate the contraction of frog sartorius muscle (Wilkie, 1956; Jewell and Wilkie, 1958). The sartorius is a parallel-fibred striated muscle in the leg. A light lever is pivoted in the middle. A weight hangs from it, close to the pivot (just to the left of it in the diagram). The muscle has one end fixed and the other attached to the right-hand end of the lever, so that when it contracts it raises the weight. An adjustable stop at the left end of the lever prevents the weight from stretching the muscle when it is not contracting. Movements of the lever are recorded by a device involving a photoelectric cell connected to a cathode ray oscilloscope (a similar device is explained on page 293). The muscle is made to contract by a series of electric shocks applied through the electrodes which are marked $+$ and $-$ in the diagram.

One point needs explaining. It is required to have the force on the contracting muscle as nearly constant as possible, to simplify analysis of the results. The force will not be constant, but will increase when the weight and lever are accelerated. The increases can be minimized by using a light lever, and by using a heavy weight hanging close to the pivot rather than a lighter one hanging further from it. A weight of mass m at a distance r from the pivot will exert a moment mgr about the pivot and

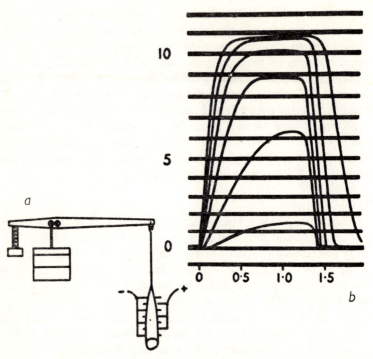

Figure 43. (*a*) Apparatus used to study the contraction of striated muscle. (*b*) The results of an experiment with the apparatus. The curves are graphs of muscle shortening (in millimetres) against time (in seconds) obtained with the same frog sartorius muscle contracting against different forces. The forces were, in order, 1 g wgt (top curve), 2·5, 5, 10, 20 and 30 g wgt (bottom curve). (From Wilkie, 1956)

have moment of inertia mr^2 about the pivot. A given moment can be obtained with least moment of inertia, by making r small. The less the moment of inertia of the weight, the less the force on the muscle will change when the lever is accelerated.

Fig. 43*b* shows a set of results obtained with the apparatus. The curves are graphs of lever movement against time, as they appeared on the screen of a cathode ray oscilloscope. They show contractions against various forces. The muscle contracted against each force until it reached an equilibrium length at which it remained until the stimuli were stopped after about one second. The smaller the force, the further the muscle contracted. The maximum force the muscle can develop falls off as it shortens, and this sets a limit to the amount the muscle can shorten against any given force. The maximum force at the optimum length is about 2×10^6 dyn/cm² of muscle cross-section.

Not only does the muscle contract further when the force is small but it also contracts faster. The experimental results can be fitted very closely by an equation devised by Hill (1938) and expressed in a more general form by Abbott and Wilkie (1953). If a contracting muscle has length l at time t, its rate of shortening ($- dl/dt$, negative because length is decreasing) is given by the following equation. F is the force against which it is contracting, F_1 is the maximum force which can be developed at the length at which the rate of shortening was measured, and a and b are constants.

$$- dl/dt = (F_1 - F)b/(F + a) \tag{13}$$

This equation, known as Hill's equation, has been shown to apply very accurately to a great variety of muscles from both vertebrates and invertebrates. The constant a is about 4×10^5 dyn/cm² of muscle cross-section but the constant b varies from muscle to muscle (Hill, 1956). Note that even when there is no force resisting contraction, the muscle contracts at a limited rate. When $F = 0$, $-(dl/dt) = F_1 b/a$. Hill's equation cannot be explained simply as a consequence of the viscosity of the muscle, but there have been attempts to explain it in other ways (Worthington, 1962; Caplan, 1966).

Another type of experiment is shown in Fig. 44a. The apparatus is the same as before except that a stop is arranged at the right hand end of the lever which prevents the muscle from shortening, until it is removed by means of a solenoid. The muscle is stimulated and builds up tension until the stop is suddenly withdrawn. There is a very quick immediate contraction, followed by slower contraction (Fig. 44b). The slower part of the contraction is just like the contractions recorded in the previous experiments (Fig. 43), and its rate depends on the load in the same way. The initial very quick contraction seems to be an elastic recoil. It seems that the muscle, prevented from shortening, stretches something elastic which recoils rapidly

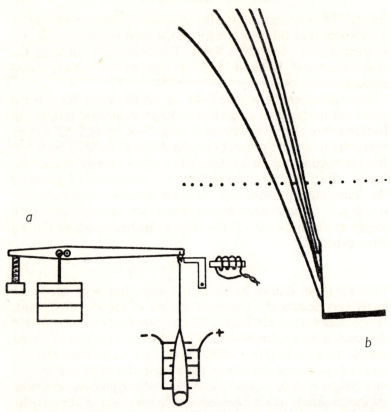

Figure 44. (*a*) Apparatus used to investigate the series elastic component of striated muscle. (*b*) The results of an experiment with the apparatus. The curves are graphs of shortening against time (which reads from right to left) obtained with the same frog sartorius muscle contracting against different forces. The dots mark 20 ms intervals. (From Wilkie, 1956)

when the stop is withdrawn. This something is known as the series elastic component.

The results shown in Fig. 44*b* were obtained with a muscle which was released at a particular time after the start of stimulation. It was known from other experiments that it would then be exerting a certain force, F_t. When the stop was withdrawn, there was only the force F, due to the weight, to resist its contraction. Hence, the elastic recoil was due to a reduction in force $(F_t - F)$, and the stiffness of the series elastic component can be calculated from the experimental results. A correction must of course be made for the elasticity of the apparatus. It seems that only part of the series elastic component can be explained as due to the elasticity of the tendons, and that part must be due to elasticity within the muscle tissue itself.

If a muscle is made to contract with its ends fixed a certain distance apart, the maximum force it develops depends on the distance between the fixed points. It diminishes as the distance is reduced below the resting length of the muscle; one would expect this, from the results shown in Fig. 43. The force of contraction also diminishes if the distance is increased above the resting length. By the force of contraction, I mean the difference between the total force exerted by the stimulated muscle, and the elastic restoring force due to its having been stretched beyond its normal length. The most revealing experiments on the relationship of force to length, have been done on isolated striated muscle fibres (Edman, 1966; Gordon, Huxley and Julian, 1966).

The striations on a muscle fibre move further apart when the fibre is stretched, and closer together when it contracts. Fig. 45 is a graph of the force of contraction of the fibre, against the distance apart of successive striations. The striations are $2 \cdot 1 \mu$ apart in slack, resting fibres ($1 \mu = 10^{-4}$ cm). The force of con-

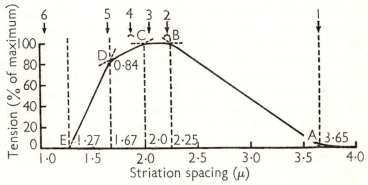

Figure 45. A graph of the force of contraction of a striated muscle fibre against the spacing of its striations. (From Gordon, Huxley and Julian, 1966)

traction is greatest when the striations are $2\cdot0$–$2\cdot2$ μ apart, and this force is taken as 100%. It is zero when the striations are $1\cdot3$ μ apart and when they are $3\cdot7$ μ apart. This can be explained in terms of the sliding filament hypothesis of muscular contraction.

A striated muscle fibre is a cell which contains many fibrils which are themselves striated. Fig. 46 is a diagram based on electron micrographs of muscle, and shows the structure of a fibril. The fibril consists of interdigitating filaments of the proteins actin and myosin running parallel to its length. They are arranged in a pattern which repeats along the fibre and is plainly the basis of the striations seen with the light microscope.

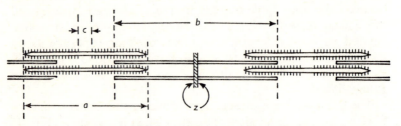

Figure 46. A diagram which is explained in the text showing the arrangement of the filaments in a striated muscle fibre. (From Gordon, Huxley and Julian, 1966)

The actin filaments are the thinner ones, whose length is indicated by the dimension b in Fig. 46. They pass through partitions known as the Z discs, which run right across the fibril. The myosin fibres (length a) are thicker and have lateral projections which attach to the actin filaments, forming bridges which are believed to be responsible for the force of contraction. It is not known how they exert this force, but suggestions have been made (Davies, 1963; Ingels and Thompson, 1966). There is a region (c) in the middle of each myosin filament which has no lateral projections. When the muscle contracts or is stretched, the actin and myosin filaments slide past each other, so as to overlap more or less.

Fig. 47 shows the extent to which the filaments overlap when the distance from one Z disc to the next (i.e. the striation spacing) has various values. The striation spacings for the

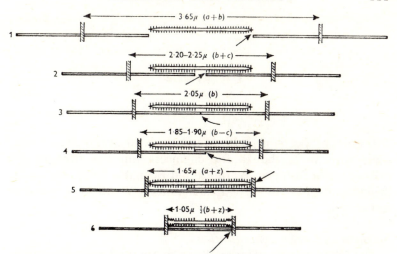

Figure 47. Diagrams showing how much the two types of filament in a striated muscle fibre overlap, at various striation spacings. (From Gordon, Huxley and Julian, 1966)

positions 1 to 6 in this figure are also indicated by numbered arrows at the top of Fig. 45. When the spacing is $3 \cdot 65 \ \mu$ (position 1), the actin and myosin filaments just fail to overlap, and one would not expect any force of contraction. It is about at this spacing that the force becomes zero in a stretched muscle fibre. As the spacing is reduced, the actin filaments overlap more and more with the myosin filaments until at a spacing of $2 \cdot 2 \ \mu$ (position 2) every projection on the myosin filaments has an actin filament beside it. If the bridges are responsible for the force, one would expect the force to be proportional to the amount of overlap in this range, and this is found to be the case. As shortening continues beyond position 2, the number of bridges which can be formed remains constant, and so does the force, until the spacing reaches $2 \cdot 05 \ \mu$ (position 3). At this point actin filaments meet end to end, and the force begins to decline. It goes on declining slowly until the spacing reaches $1 \cdot 65 \ \mu$ (position 5) when the ends of the myosin filaments come into contact with the Z discs. Further contraction must crumple the myosin filaments, and the force declines more sharply beyond this point, until it disappears altogether.

CHAPTER 4

STRENGTH

THE parts of mechanics which form the basis of this chapter are dealt with in textbooks for engineers, on the strength of materials (for instance, Warnock and Benham, 1965). They will probably be new to most readers, so they will be explained rather fully.

Strength and stress concentration

The two measures of strength which are used most are tensile strength and compressive strength. Tensile strength refers to tests in which the material is stretched and compressive strength to tests in which it is compressed. In each case the strength is the force which is just sufficient to break a specimen of the material divided by the initial cross-sectional area of the specimen. The area is measured where the specimen breaks in a plane at right angles to the direction of the forces. Notice that it is the initial cross-sectional area, before stress is applied, that is used. We saw on page 68 that there are two ways of defining tensile or compressive stress. It can be defined in terms of the initial area A or of the stressed area $(A - \Delta A)$. The former definition is always used in discussions of strength though the latter one is often preferred in discussions of elasticity and viscosity. The difference is only important for materials like rubber which can be stretched or compressed a long way before they break.

The definitions of strength assume that the specimen is evenly loaded, so that equal parts of its cross-sectional area withstand equal forces. A break may start anywhere where the force per unit area exceeds the strength, even if the average force per unit area across the whole specimen is well below the strength.

When tensile strength is being measured, the ends of the specimen must be gripped in some way, for instance, by clamps. It is very hard to say how the total force is distributed across a cross-section close to a clamp, and if the specimen breaks near a clamp it is hard to estimate the strength. It is therefore best to use specimens shaped so that they will always break well away from the clamps, where the force is evenly distributed across the cross-section. Engineers normally use specimens which are wide at the ends and taper fairly gently to a slender middle section (Fig. 48a). The clamps are fixed to the large ends and the specimen can be relied on to break across the middle. Ring-shaped specimens are also suitable (Ritchie, 1965).

There is no need for clamps when compressive strength is being measured, and a cylinder or rectangular block of material is quite suitable. The ends of the specimen and the surfaces

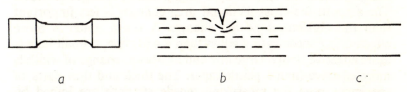

a   *b*   *c*

Figure 48. (*a*) A sample of material of a suitable shape for the measurement of tensile strength. (*b*), (*c*) Diagrams illustrating the account of stress concentration

compressing it should be flat and parallel. The length of the specimen, measured in the direction of the forces, should not be very much more than its other dimensions, lest the specimen buckle instead of being simply crushed. We shall discuss buckling later (page 155).

Some materials yield before they break. As stress increases, they behave elastically up to a certain limit. If the stress is increased beyond this elastic limit, the material flows and no longer returns to its initial dimensions when the stress is removed. Such materials are described as tough, and breakage preceded by obvious yielding is called tough fracture. Brittle materials break abruptly without yielding. A material may be tough in some conditions and brittle in others; for instance, rubber is normally tough and glass is normally brittle, but rubber becomes brittle at very low temperatures and glass becomes tough at high ones.

An account of stress concentration will show one reason why

it is often much easier to break a brittle material than a tough one (another reason will appear on page 121). Fig. 48*b* shows a bar with a notch in it being stretched. Each of the lines along it represents an equal part of the total force. Well away from the notch, the force is distributed uniformly over the cross-section of the bar. This has been represented in the diagram by spacing the lines evenly. Near the notch, the distribution of force is disturbed. The stress is greater close to the tip of the notch than it is elsewhere, as the closely spaced lines indicate. This situation is known as stress concentration. The notch of course reduces the cross-sectional area of the bar, but a very shallow notch can have a very large stress concentration at its tip. Stress concentrations are also set up by holes and by changes of width. The amount the cross-sectional area changes is less important than the abruptness of the change. A sharp notch is more effective in concentrating stress than a wide one, a small hole is more effective than a large one and an abrupt change of width is more effective than a gentle taper. The thick and thin parts of specimens used for measuring tensile strength are joined by gentle curves to avoid serious stress concentration (Fig. 48*a*). Peterson (1953) gives a lot of quantitative information about stress concentrations. The stress close to a small round hole running transversely through a bar, which is being stretched lengthwise, is three times the average stress in the bar. Cracks are, in effect, very sharp notches, and the stress close to the end of a crack may be very many times the average stress in the specimen.

Consider what happens when a brittle bar with a notch in it is stretched. For any applied force, the stress at the tip of the notch will be greater than the stress elsewhere in the bar. As the force increases, a time will come when the stress at the tip of the notch exceeds the tensile strength although the average stress across the whole specimen is less than the tensile strength. A crack will form at the tip of the notch. The crack acts as a very sharp notch, giving rise to high stress concentrations, and it therefore spreads across the bar. The bar breaks, although the average stress across its cross-section was less than the tensile strength. A small notch can make a brittle specimen very much weaker. A notched bar (Fig. 48*b*) is weaker than a bar without a notch whose cross-sectional area equals its minimum cross-sectional area (Fig. 48c) if the material is brittle.

A notched bar of tough material would behave differently. When the stress at the apex of the notch reached the elastic limit, the material there would yield and the stress concentration would disappear. Notches and other irregularities which cause stress concentrations do not reduce strength in tough materials. If the bars shown in Fig. 48b, c were made of tough material, they would be about equally strong. However, the notched specimen would be more susceptible to fatigue.

If a piece of material suffers fluctuating stresses instead of a steady stress, and if the fluctuations continue for long enough, the material may break, even though the maximum stress is well below its strength. This phenomenon is called fatigue.

Bone

Bone must obviously be strong to serve its functions. A lot of people have investigated its strength but their findings will mean more to us if we examine first the composition and structure of bone.

About two thirds of the weight of bone, or half its volume, is inorganic material known as bone salt. It contains calcium, phosphate and hydroxyl ions in proportions which correspond fairly closely to the formula of hydroxyapatite, $3Ca_3(PO_4)_2.Ca(OH)_2$, with small quantities of other ions. It is present as tiny crystals often about 200 Å long (Bourne, 1956). The rest of bone is organic material, and this is mainly collagen. The crystals of bone salt lie between the collagen fibrils and are believed to be attached firmly to them (Marino and Becker, 1967). The long axes of the crystals are parallel to those of the fibrils.

Groups of collagen fibrils run parallel to each other to form fibres in the usual way. These fibres are arranged differently in different types of bone. Smith (1960a, b) has described the arrangements found in mammals. In woven-fibred bone the fibres are tangled. In the other types the fibres are laid down neatly in lamellae. The fibres in any one lamella are parallel to each other but the fibres in successive lamellae are about at right angles to each other. There are two main types of bone with this sort of structure (Fig. 49). One is surface bone, which has the lamellae parallel to the surface of the bone. The other type consists of osteones. Each osteone has a central canal containing blood vessels with lamellae arranged concentrically round it. The collagen fibres run helically in the lamellae but are often more or less parallel to the canal in one lamella and more or less at right angles to it in the next, so that they can be thought of as longitudinal and circumferential. There are often more longitudinal fibres than circumferential ones. In long bones the osteones run parallel to the long axis of the bone.

Osteone bone may be compact or spongy. Compact bone is solid, apart from the canals for blood vessels, cavities for cells and the very fine canals known as canaliculi which radiate from

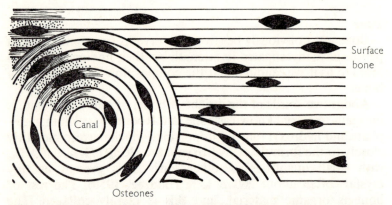

Canal

Surface
bone

Osteones

Figure 49. A schematic section through part of a mammal bone, showing surface bone and osteones. The directions of the collagen fibres are indicated in some of the lamellae. Cell cavities are shown black but canaliculi are omitted

the cell cavities and connect them together. Spongy bone is, as its name indicates, spongy. Its sponginess can be seen with the naked eye.

Bone can be turned on lathes or shaped in milling machines. Various investigators have machined small pieces of suitable shape from large bones, and have measured their tensile and compressive strengths in machines designed for testing engineering materials. The bone should not be allowed to get too hot during machining, lest heat affect its properties. The bones should be tested in their natural wet condition, for dried bones have different properties. From experiments with fresh mammal bones it seems that Young's modulus is about 10^{11} dyn/cm^2 (Smith and Walmsley, 1959), the tensile strength is about 10^9 dyn/cm^2 (Evans, 1957; Currey, 1959) and the compressive strength is about $1 \cdot 8 \times 10^9$ dyn/cm^2 (Evans, 1957). All these values are for compact bone, stressed parallel to the osteones. Similar values for Young's modulus and tensile strength have been obtained in experiments using single osteones dissected from bones, instead of bigger chunks of bone (Ascenzi, Bonucci and Checcucci, 1966). The tensile strength seems to be less when the bone is stressed at right angles to the osteones (Evans, 1957).

Bone is much stiffer and less extensible than the collagen it contains. Young's modulus for collagen from decalcified bone is about 10^{10} dyn/cm^2 (Ascenzi, et al., 1966), or the same as for collagen from other tissues, but Young's modulus for bone containing 50% of collagen by volume is about 10^{11} dyn/cm^2. The difference must be due to the bone salt, and one would expect the bone salt to raise the modulus of bone if the crystals were attached to the collagen fibres or to each other, since the modulus of its crystals is probably of the order of 10^{12} dyn/cm^2 (Currey, 1964).

The bone salt apparently makes the bone stronger as well as stiffer. Ascenzi and his colleagues found that the tensile strength of collagen from bone is about 9×10^8 dyn/cm^2 (about the same as for tendon collagen), while the tensile strength of bone was about $1 \cdot 3 \times 10^9$ dyn/cm^2 although only half its volume is occupied by collagen.

Bone salt does not at first sight seem a promising material for making bone strong, since it is presumably brittle like other crystalline salts. Currey (1962b, 1964) has suggested that the properties of bone can be understood by comparing it with composite materials used in engineering.

Bone and other composite materials

Composite materials are solids consisting of two materials which are mixed together but retain their separate identity. A great many are used in industry and engineering and have been duly studied (Holliday, 1966). It is only necessary to consider two of them which seem to throw light on the nature of bone. They are fibreglass and filled rubber.

Fibreglass consists of fine fibres of glass embedded in a plastic such as a polyester resin. It is used for making many things including fishing rods and boats. Its value lies in its being both strong and light. The resin has a relatively low Young's modulus and the glass a relatively high one, and the modulus of the fibreglass lies between these. An object made of fibreglass is stronger than it would be if it were made of the resin alone, or if it were made of solid glass. Fibreglass owes its strength both to the strange fact that fine, undamaged glass fibres have a much higher tensile strength than thicker pieces of glass, and to the low modulus of the resin. It is not clear why glass fibres should be so strong, but it has been claimed that they are relatively free from flaws that set up stress concentrations in thicker pieces of glass. Once a crack starts in glass, it spreads easily because of the stress concentration at its end. If a crack starts in solid glass the whole piece will probably break. If a crack starts in a glass fibre in fibreglass, it will probably go right through the fibre but it will stop when it gets to the resin, because the resin will stretch and relieve the stress concentration.

Currey (1962b, 1964) compared the resin in fibreglass to the collagen in bone, and the glass to the bone salt. Bone, like fibreglass, has a Young's modulus between the moduli of its constituents and a tensile strength which is probably greater than the strength of a large piece of either constituent. However, there is an important difference between fibreglass and bone. The strength of fibreglass depends on the glass fibres being reasonably long, though they may be much shorter than the piece of fibreglass. Fibres less than about 1 cm long are un-satisfactory (Holliday, 1966). The crystals in bone are only about 200 Å long, and while they may be bonded together in some way there is no indication of the existence of long strong units.

Filled rubbers perhaps offer a better analogy to bone, though they are neither as strong nor as stiff. They are mixtures of rubber with fine particles of another material which is usually carbon black. Carbon black is fine soot. Its particles are roughly spherical and in some of the grades used for reinforcing rubber are about the size of the crystals in bone. Rubber filled with carbon black is used for making tyres.

The rubber seems to become very firmly attached to the carbon black, and the attachments prevent the molecules from straightening as much as they otherwise would, when the rubber is stretched. This is particularly so since the carbon particles tend to join together in short chains (Holliday, 1966). Hence, Young's modulus is higher than for pure rubber. The tensile strength is also higher. Rubber containing 50% carbon black can be as much as sixteen times as strong as pure rubber (Bueche, 1958).

Stress concentrations in bone

Bone has channels for blood vessels and (except in some fish) cavities for cells. Currey (1962a) has considered whether these are likely to weaken it seriously, by acting as stress concentrators.

A long bone is probably in most danger of being broken by forces which tend to bend it. If the channels for blood vessels ran transversely through bones, the stress near them when the bone was bent would be concentrated by a factor of 3 (Peterson, 1953). In fact, most of them run more or less parallel to the long axes of the bones and can have little or no stress concentrating effect when the bone is bent. Only a few anastomosing channels run at any great angle to the axis of the bone.

The cavities for the bone cells in surface bone and osteone bone are (roughly) oblate spheroids: that is, they have the shape of a squashed sphere. Their stress concentrating effect will depend both on how flattened they are, and on their orientation. Currey (1962a) measured some of them in human and cattle bones and concluded that most of them would concentrate the stress by a factor of about 4, if they were oriented in the least favourable way. In fact, they are generally oriented in the most favourable way, with their short axes at right angles to the long axis of the bone (Fig. 49), and they can have little stress concentrating effects.

Currey suggested that the bone cell cavities might actually strengthen bone, by helping to prevent cracks from spreading.

The stress concentration at the end of a crack may be enormous, because the crack is a very sharp notch. When a crack runs into a bone cell cavity it is in effect blunted. The principle is used by makers of light plastic raincoats: the pockets of these coats open through slits in the plastic but a round hole is cut at each end of the slit to reduce stress concentrations and the danger of tearing.

Currey tested his idea. He made small blocks of bone and cracked them with a chisel. Many broke in two and were rejected but some cracked only part of the way across. He made sections of these and looked at the end of the crack in each section, under a microscope. In some sections the crack ended in solid bone, but cracks stopped in bone cell cavities far more often than if they were stopping at random. Naturally some sections must show cracks stopping in bone, since there are gaps between one cavity and the next.

The canals for blood vessels may also help to stop cracks but will be less effective than the bone cell cavities, because they are further apart.

Echinoderm ossicles

A great many animal skeletons contain crystals of inorganic salts. In bone, mollusc shells, corals, etc., the crystals are small and are separated by protein. These composite materials are presumably all stronger than if they consisted of solid pieces of the pure salt (see page 117). The ossicles of echinoderms are exceptional in being single crystals of calcium carbonate. They are not compact crystals, but three-dimensional networks of bars of calcium carbonate. The network structure may make them stronger than they would otherwise be. A crack in a compact crystal would probably spread right through the crystal but a new crack must be formed for every bar if the network is to break (Currey and Nichols, 1967).

Insect cuticle

Insect cuticle consists mainly of chitin and protein. Chitin is a polysaccharide, which differs from cellulose only in having one $-OH$ group in each glucose unit replaced by $-NHCOCH_3$. It is a fibrous material, and its fibrils are laid down in lamellae. The protein seems to cement the fibrils together. In the exocuticle, which is the layer that gives the stiff parts of insect cuticle their stiffness, the protein is heavily cross-linked by the chemical process known as quinone tanning (Pryor, 1962; Brunet, 1967).

Jensen and Weis-Fogh (1962) stretched locust tibias, and so measured the Young's modulus and tensile strength of the cuticle. Their results show that the properties of the cuticle are very like the properties of bone and oak (Table 3, page 72).

Impact strength

The work done in deforming a piece of material which obeys Hooke's law through a distance x is $Fx/2$, where F is the elastic restoring force when the deformation is complete (page 71). We can calculate the amount of work needed to break a piece of material.

It will be convenient to consider a piece of material being stretched, though we might equally well have considered compression, bending or any other mode of deformation. Let the piece have length l and cross-sectional area A when it is undeformed. Let its tensile strength be T and Young's modulus E. The force needed to break it is AT. If we can ignore the change in cross-sectional area due to stretching, the amount this force will stretch the specimen is Tl/E. The work done in stretching the specimen till it breaks is therefore $AT(Tl/E)/2$ or $AlT^2/2E$. Al is of course the volume of the specimen, so the energy needed to break it can be given as $T^2/2E$ per unit volume of material.

This will only be true if the material obeys Hooke's law right up to the moment when it breaks. If the material is tough, and yields before it breaks, more work will be needed.

The energy needed to break a unit volume of material is known as the impact strength of the material. In many situations it is the critical property which decides whether a specimen will break or not. A cup which is dropped on the floor cannot break unless its kinetic energy, at the moment of striking the floor, is sufficient to do the work of breaking it.

Notice that the impact strength, $T^2/2E$, is proportional to the square of tensile strength and inversely proportional to Young's modulus. The higher the tensile strength and the lower the modulus, the greater is the impact strength. Holes, notches and other irregularities which concentrate stresses make specimens much easier to break by impact. For instance, a small transverse hole through a brittle bar will in effect reduce the tensile strength by a factor of 3, and so it will reduce the impact strength (which is proportional to the square of the tensile strength) by a factor of 9. Padding makes a specimen harder to break by impact. A cup is less likely to break if it is dropped on a carpet than if it is dropped on concrete, because some of its kinetic energy is used in deforming the carpet.

The impact strength of a visco-elastic material depends on the speed at which it is deformed. If it is deformed slowly, it will deform more before breaking and more energy will be needed to break it, than if it had been struck a sudden blow.

Various machines are made for measuring impact strength. In the most popular ones, a specimen of the material is bent and broken by a fast blow from a heavy pendulum. Fig. 50 illustrates the principle of these machines. The pendulum of mass m is swung up to a height h_1, so that it has potential energy mgh_1. It is released, and strikes the specimen at the bottom of its swing. It breaks the specimen and swings on, rising to a height h_2 where it has potential energy mgh_2. If energy losses due to friction, etc., can be ignored, the energy used in breaking the specimen is $mg(h_1 - h_2)$. It is usual to cut a notch in specimens

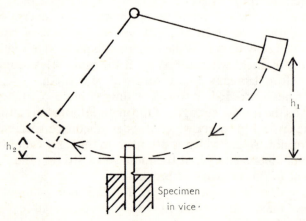

Figure 50. A diagram illustrating a technique for measuring impact strength

which are to be tested in machines of this kind at the place where the break will start. A notch makes the specimen more likely to break in brittle fashion and reduces the scatter of results (Ritchie, 1965).

Impact strength depends so much on the precise conditions of testing that it is very hard to predict the impacts a material will stand in other conditions, from the results of tests.

Protection of bones by skin

The bones of animals are padded by skin and other tissues which must help to protect them from impacts. Currey (1968) has investigated the protection which the skin gives to the metatarsal bones of rabbits. He used a machine which is like the one represented in Fig. 50, except that the specimen is held at both ends and struck in the middle. He used complete metatarsals as the test specimens, and did not notch them. In some of the tests the bone was left bare but in others it was padded with a piece of the furry skin that covers the metatarsals when the foot is intact. The skin was simply set against the bone, where the hammer would strike it. The average energy needed to break the padded bones was 37% more than the average energy needed to break the bare ones. The skin apparently gives quite a useful degree of protection.

Energy stored for a flea's jump

If a material has impact strength H and resilience (in the sense defined on page 70) R, the maximum amount of work that can be obtained from a unit volume of it in an elastic recoil is $HR/100$.

Before fleas jump, they store elastic energy in pieces of resilin at the bases of their hind legs (page 80). Elastic recoil of the resilin seems to provide the energy for the jump. Is there enough resilin to store the energy that is needed? This question has been discussed by Bennet-Clark and Lucey (1967).

The flea *Spilopsyllus* weighs about 0·45 mg and leaves the ground, when it jumps, at about 100 cm/s. Its kinetic energy must be $\frac{1}{2} \times 0·45 \times 10^{-3} \times 100^2 = 2·25$ erg. This is the amount of energy which must be stored by the resilin. The impact strength of resilin cannot be calculated from the formula given on page 121 because resilin does not obey Hooke's law at large extensions, but it can be estimated from the area under graphs of force against extension, for pieces of resilin which were stretched till they broke (Weis-Fogh, 1961b). It seems to be about 2×10^7 erg/c³. The flea has two pieces of resilin whose total volume is about 3×10^{-7} c³. Hence, the work needed to break both pieces of resilin would be 6 erg. If they were deformed almost to the breaking point and allowed to recoil, nearly 6 erg could be obtained from them since the resilience of resilin is almost 100% (Jensen and Weis-Fogh, 1962: the recoil when a flea jumps is rather faster than in any of the experiments in which resilience was measured, but not so much faster as to be likely to affect the resilience very much). Almost 6 erg could apparently be obtained in the recoil, and only 2·25 erg is needed for the jump. The hypothesis that the energy for the jump is stored in the resilin is, therefore, feasible.

Bending

Fig. 51a shows a beam with one end rigidly fixed to a wall. In Fig. 51b a couple of moment M is acting on the free end of the beam. This has bent the beam, so that it forms an arc of a

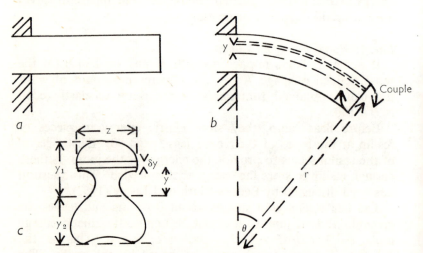

Figure 51. Diagrams illustrating the discussion of a beam fixed at one end, bent by a couple acting on the other end

circle. Bending has stretched the upper parts of the beam and compressed the lower parts, but there is an intermediate layer in the beam which is neither extended nor compressed. It is of course infinitely thin, and is better described as a surface than as a layer. It is known as the neutral surface. Though it is neither stretched nor compressed it is bent. It is represented in Fig. 51b by the broken line whose radius of curvature is r.

Consider the thin layer of material, of thickness δy, which lies at a distance y from the neutral surface, towards the outside of the curve. It was originally the same length as the neutral surface but has been stretched and is now $(r + y)/r$ times as long. In other words, it has suffered a tensile strain y/r. The stress needed to produce this strain is Ey/r, where E is Young's modulus for the material of the beam. We could of course have considered a layer on the inside of the curve from the neutral axis. The distance y would then be negative and the stress would be a negative tensile stress, which is a compressive stress.

Fig. 51c shows the beam in end view. It is represented as having an irregular but symmetrical cross-section. The width of the thin layer we have been considering is z, so its cross-sectional area is $z\,\delta y$. Since the stress is Ey/r the force stretching the layer must be $Eyz\,\delta y/r$. The total force acting on the end of the beam could be got by dividing the beam into a large number of thin layers, calculating the force for each, and adding them together. Since there is only a couple acting on the end of the beam, and no net force, they must add up to zero. The total force stretching the upper part of the beam must equal the total force compressing the lower part.

In this way one can show where the neutral surface must be. It turns out that the neutral surface of a cylindrical beam passes through the axis of the cylinder, and the neutral surface of a beam of rectangular section is half way through its thickness. There is a very simple method of finding the neutral surface for beams of other shapes. Cut out a piece of cardboard of the same shape as the cross-section and find its centre of gravity (the method described on page 20 is suitable). The neutral surface must pass through the position corresponding to this centre of gravity on the actual cross-section.

The beam is being bent by a couple of moment M. Not only

must the forces stretching the upper part of the beam equal the forces compressing the lower part, but the two sets of forces must form a couple of moment M. It will be convenient to take moments about the neutral axis of a cross-section of the beam. This is the line where the neutral surface cuts the cross-section. The force $Eyz\ \delta y/r$ which acts in the thin layer of the beam which we have been considering, has a moment $Ey^2z\ \delta y/r$. The moment M is the sum of the moments of the forces on all the individual layers

$$M = \sum(Ey^2z\ \delta y/r)$$
$$= (E/r)\sum(y^2z\ \delta y)$$

The expression $\sum(y^2z\ \delta y)$ is known as the second moment of area of the cross-section and is usually represented by the letter I so that we can write

$$M = EI/r \qquad (14)$$

The greater I is, the greater will r be for given values of M and E. If r is large the amount of bending is small. A large value of I makes a beam hard to bend. The second moment of area of a cross-section can be obtained by drawing an outline of the cross-section, dividing it up into narrow strips parallel to the neutral axis, measuring each strip and adding together the values of $y^2z\ \delta y$. Some values are given in Fig. 52.

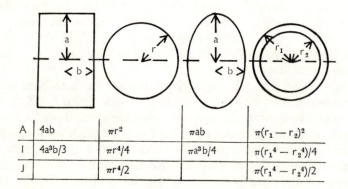

A	$4ab$	πr^2	πab	$\pi(r_1 - r_2)^2$
I	$4a^3b/3$	$\pi r^4/4$	$\pi a^3b/4$	$\pi(r_1{}^4 - r_2{}^4)/4$
J		$\pi r^4/2$		$\pi(r_1{}^4 - r_2{}^4)/2$

Figure 52. Cross-sections of rectangular, cylindrical, elliptical and tubular beams with expressions for cross-sectional area (*A*), second moment of area for bending in a vertical plane (*I*, the neutral axes are indicated by broken lines) and second moment of area for twisting (*J*)

We saw that the stress at a distance y from the neutral surface of the beam is Ey/r. From equation (14), this is equal to My/I. Suppose the cross-section of the beam extends to a distance y_1 above the neutral surface and y_2 below it (Fig. 51). The upper parts of the beam are stretched and the greatest tensile stress is at the top surface of the beam. It is My_1/I. Similarly the greatest compressive stress is My_2/I, at the bottom surface of the beam. The beam will break if My_1/I exceeds the tensile strength of the material or if My_2/I exceeds the compressive strength.

The beam we have been considering (Fig. 51) had a couple acting on its end, and bent into an even curve. Fig. 53a shows a

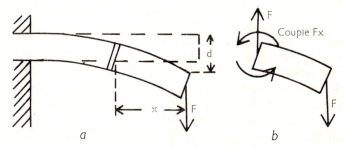

Figure 53. Diagrams illustrating the discussion of a beam fixed at one end, bent by a single force acting on the other end

beam with a single force, not a couple, acting on its end. Consider the effects of the force F on the thin slice of beam at a distance x from it. They are, perhaps, best appreciated by thinking of the beam cut through just beyond the slice. What forces would be needed at the cut surface to keep the piece of beam cut off in this way in equilibrium (Fig. 52b)? Plainly, an upward force F and an anticlockwise couple of moment Fx are needed. Therefore, when the beam is intact, a downward force F and a clockwise couple of moment Fx must act on the outer surface of our thin slice. The slice will be both sheared by the force and bent by the couple. If the cross-section has area A, there will be a shearing stress F/A. If the second moment of area about the neutral axis is I there will be a tensile or compressive stress Fxy/I at a distance y from the neutral axis. The shearing stress is constant along the beam but the tensile and compressive stresses increase as x increases. If the beam is short and fat the tensile and compressive stresses will be relatively small, even at the fixed end, and it will shear without bending much. If it is long and thin the tensile and compressive stresses will be relatively high, and bending will be more important than shearing.

Suppose the cross-section of the beam is symmetrical and uniform all along the beam. We wish to know the distance d that the force will pull down the end of the beam (Fig. 53a). If the beam is sufficiently short and fat we can ignore bending and calculate d from the shear modulus (page 69). If it is sufficiently long and thin, we can ignore shear. It can be shown that if this is the case and if d is small relative to the length l of the beam

$$d = Fl^3/3EI \qquad (15)$$

where E is Young's modulus (see for instance Warnock and Benham, 1965). Similar formulae can be obtained for other arrangements of beams, such as beams with loads spread out along their length and beams supported at both ends. These formulae can be found in textbooks on the strength of materials (e.g. Warnock and Benham, 1965).

Strength with lightness is often desirable, both in engineering and in animal skeletons. How is it best achieved with a given material? The problem is simple enough if the member has only to resist tension along its length, or compression if there is no danger of buckling (see page 155). Suppose a cable of given length is wanted to suspend a given load. The strength of the cable will be the strength of its thinnest part, so the cable can be made lightest if it is made uniform, with the same cross-sectional area all along. The shape of the cross-section will be of no consequence since strength in tension is simply proportional to cross-sectional area. The design of members which need strength in bending is more complicated. Suppose a horizontal beam of given length is required to support a given load. It should not be made uniform, but should be thickest where the bending moments are greatest, if lightness is important. The beam in Fig. 53 could be made lighter without weakening it, by tapering it towards the free end. At the extreme end where there is no bending moment, it need only be thick enough to withstand the shearing stress. The shape of the cross-section of a beam has a great bearing on its strength in bending. We will consider a few simple cross-sections and see which can be made lightest to withstand a given bending moment.

A beam which is being bent will break if the maximum tensile stress My_1/I (see above) exceeds the tensile strength T (i.e. if $M > TI/y_1$) or if the maximum compressive stress My_2/I exceeds the compressive strength C (i.e. if $M > CI/y_2$). We will estimate the cross-sectional areas needed to withstand a bending moment M for beams of various shapes. We will suppose that the beam breaks first where it is in tension rather than where it is compressed, but the only effect of this supposition is to make us use the symbols y_1 and T where we would otherwise have used y_2 and C.

Consider first a cylindrical beam of radius r. $I = \pi r^4/4$ (Fig. 52) and $y_1 = r$ so $TI/y_1 = T\pi r^3/4$. By solving the equation

$$M = T\pi r^3/4$$

we find

$$r = (4M/\pi T)^{1/3}$$

and the cross-sectional area, πr^2, is $\pi^{1/3}(4M/T)^{2/3}$.

Now consider a beam with an elliptical cross-section with axes a and b as shown in Fig. 52

$$M = T\pi \, a^2 b/4$$
$$a = [(a/b)4M/\pi T]^{1/3}$$

and the cross-sectional area, $\pi \, ab$, is $(\pi \, b/a)^{1/3}(4M/T)^{2/3}$. The elliptical beam will be lighter than the cylindrical one if $a > b$: that is, if the major axis of the ellipse lies in the plane of bending. The more slender the ellipse, the lighter can the beam be made, within limits. If it is made too slender the beam will tend to give way by bending sideways, owing to elastic instability (page 157) and will have to be made heavier than this calculation predicts.

Now consider a cylindrical tubular beam with external and internal radii r_1 and r_2, as shown in Fig. 52.

$$M = T\pi(r_1^4 - r_2^4)/4r_1$$
$$= T\pi r_1^3(1 - k^4)/4$$

where

$$k = (r_2/r_1)$$
$$r_1 = [4M/\pi T(1 - k^4)]^{1/3}$$

The cross-sectional area is $\pi(r_1^2 - r_2^2)$ or $\pi r_1^2(1 - k^2)$, which is $[\pi(1 - k^2)/(1 + k^2)^2]^{1/3} (4M/T)^{2/3}$. This is less than the area for the cylindrical beam. A beam can be made lighter if it is made hollow, and a relatively thin wall (a high value of k) is an advantage, within limits. If the wall is made too thin the beam will tend to fail by elastic instability (page 157) and will have to be made heavier than this calculation predicts.

Tubes are stiffer than solid beams of equal strength, as well as being lighter.

Limb bones are hollow, and so can be lighter than would be necessary if they were solid. For the shaft of the human femur, k is about $0 \cdot 5$ and the cross-sectional area is therefore only $[0 \cdot 75/(1 \cdot 25)^2]^{1/3} = 0 \cdot 78$ of the cross-sectional area of a solid bone of equal strength. Bones of smaller mammals and of birds have relatively thinner walls. For the humerus of a swan, k is about $0 \cdot 9$ and the cross-sectional area may be only $0 \cdot 38$ of the cross-sectional area of a solid bone of equal strength. A tube with so thin a wall might be liable to kink like a bent rubber tube and fail by elastic instability (page 157), but thin struts of bone run across the cavity of the humerus and must help to prevent kinking.

Throughout this discussion we have considered only beams which are symmetrical about the plane of bending. Asymmetrical beams are discussed in engineering textbooks (for instance, Warnock and Benham, 1965).

Twisting

A beam is bent by couples acting in planes parallel to its length. It is twisted by couples acting in planes at right angles to its length. Twisting shears the material. The shearing stress is greatest at the outside and zero along a central axis. The shear stress at a distance y from the axis of a cylindrical tube or shaft twisted by a couple of moment M' is $M'y/J$. J is the second moment of area about the axis and is not the same as the quantity I used in estimating bending stresses (Fig. 52).

Hollow shafts give strength with lightness in twisting as well as in bending.

Strength of a bird's wing

A bird's wing is attached to its body by the humerus, which must be strong enough not to break in flight. Pennycuick (1967) has investigated the strength of the humerus of the pigeon (*Columba*) and has considered the forces it is likely to have to withstand.

Fig. 54 shows a pigeon's wing outstretched with the humerus, radius and ulna in their natural positions. When the bird flies, upward forces act on the wings and support its weight (page 228). The total upward force or lift on each wing can be thought of as acting at a particular point, the centre of lift, just as the

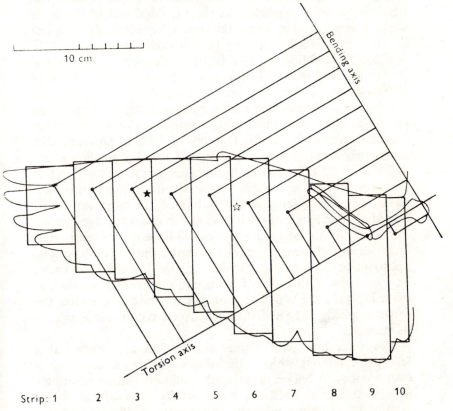

Figure 54. Outlines of a pigeon wing and some of its bones. The superimposed lines and stars are explained in the text. (From Pennycuick, 1967)

weight of a body can be thought of as acting at the centre of gravity. We see presently (page 235) how the position of the centre of lift can be estimated, using the rectangles which are superimposed on the outline of the wing in the figure. A pigeon can glide, it can fly actively forward, and it can hover momentarily. The position of the centre of lift depends, as we shall see, on which it is doing. When the bird is gliding the centre of lift will probably be close to the position indicated by the hollow star, when the bird is hovering it will probably be close to the position indicated by the solid star and when the bird is flying forward it will be somewhere between these positions.

The lift tends both to bend the humerus and to twist it. The bending moment is greatest at the proximal end of the bone where it is the product of the lift and the perpendicular distance of the centre of lift from the line marked "bending axis" in Fig. 54. The twisting moment is the product of the lift and the perpendicular distance of the centre of lift from the long axis of the bone, marked "torsion axis" in Fig. 54.

Pennycuick devised apparatus for measuring the strength of the wing bones in bending and in twisting. The apparatus for bending is shown in Fig. 55. The ends of the bone are embedded in Wood's metal in brass cups. Wood's metal is an alloy which melts at 70°C, so bones can be embedded in it without heating them too much. The bone is fixed to the apparatus in this way, to avoid confusing stress concentrations. If the bone was simply bound to the apparatus with wire, for instance, it might be sufficiently firmly held, but there would be stress concentrations under the small areas of contact with the wire which would be apt to make the bone break under the wire. If it did break there, one would have little idea of the size of the local stress that had made it break. The Wood's metal fits the bone closely so that the forces applied to it are distributed over a considerable area of its surface.

The lower brass cup is fixed rigidly. The upper one has a lever attached to it, which is pulled sideways by a string running over a pulley. A bucket is tied to the end of the string and water is poured into it until the bone breaks. The bucket is not allowed to spill but is weighed with the water still in it, and this gives the force F which broke the bone. If the distance from the

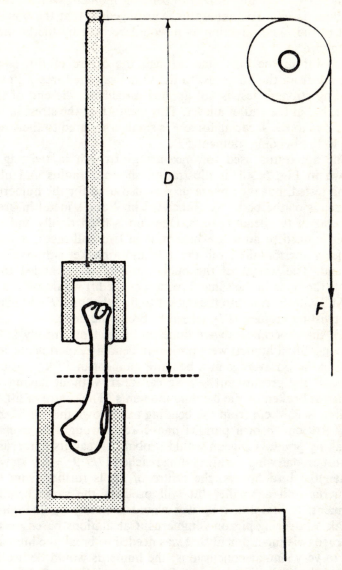

Figure 55. Apparatus for measuring the bending strength of bones. (From Pennycuick, 1967)

string to the fracture is D, the bending moment which broke the bone is FD. The apparatus is arranged so that the bone is bent in the same direction as it would be bent by the lift in a flying bird.

The lift in the bird acts through the centre of lift, some distance from the humerus. In the same way the force F in the bending strength test is not applied directly to the end of the bone, but at the end of a lever. This means that the stress in the bone due to the shearing force F is small compared to the stress due to the bending moment FD.

The apparatus used for measuring strength in twisting is shown in Fig. 56. This illustration shows a radius and ulna being tested, but the apparatus was also used for the humerus. The ends of the bone are embedded in Wood's metal in brass cups, as in the bending tests. One cup is fixed rigidly and the other is fixed to an axle which runs in two ball races. A lever projects horizontally from the axle and a bucket is hung from its end. The weight of the bucket and of water added to it twists the bone in the same direction as the lift would twist it in flight. Water is run into the bucket to find the force F and hence the twisting moment FD needed to break the bone.

All the experiments were done on bones from freshly killed pigeons. Eight humeri were broken in bending experiments, and they broke on average at a bending moment of 17 kg wgt cm. What is the greatest lift a wing can bear, without having the humerus broken by the bending moments? The centre of lift for gliding is 12·4 cm from the bending axis shown in Fig. 54, but only 9·9 cm from a parallel line drawn through the average breaking point. A pigeon would probably break its humerus if the lift on one wing during gliding reached $17/9 \cdot 9 = 1 \cdot 7$ kg wgt. When the bird hovers, the centre of lift is further from the humerus and a smaller lift will produce the same bending moment. A lift of 1·1 kg wgt would probably be enough to break a hovering pigeon's humerus. Calculations based on the average twisting moment that was needed to break the humerus led to very similar conclusions: the humerus would be broken by twisting if the lift reached 1·8 kg wgt in gliding or 1·1 kg wgt in hovering. The lifts needed to break the radius and ulna seem to be a little lower.

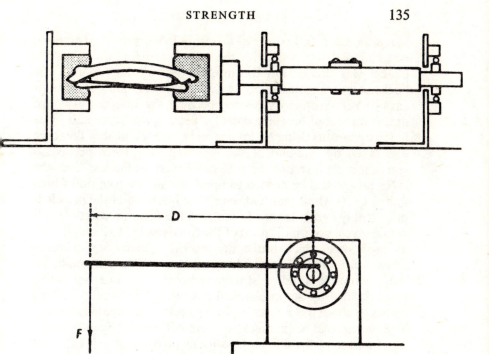

Figure 56. Apparatus for measuring the twisting strength of bones. (From Pennycuick, 1967)

In the experiments, bending and twisting moments were applied separately. In flight, they would act together. Because of this, the estimates of lifts needed to break the bones must be rather too high. We will return to this question (page 141).

Would a bone be the first thing to break if too great a lift acted on a wing, or would some other structure such as a tendon give way? The wings are flapped down in active flight and held horizontal in gliding by the pectoralis major muscle, which runs from the sternum (in the chest) to the humerus. If the lift became too great this muscle might be torn away from its insertion on the humerus. The wing would swing up passively and greater lifts could not be developed.

Pennycuick and Parker (1966) investigated the strength of the insertion of this muscle. They dissected the muscle, with the humerus still attached, from a freshly-killed pigeon. The bone was clamped in a vice with the muscle hanging down, and a bucket was hung from the muscle. Water was run into the bucket, until the muscle broke away from the bone. It was not easy to get satisfactory results, because the insertion was apt to tear instead of breaking cleanly. Tearing was presumably due to the apparatus being set up unevenly, so that most of the force was taken by one end of the insertion. The force needed to tear a muscle insertion or a piece of cloth is far less than the force that would be needed to break it with an even pull which tightened its whole width at once. The insertion of the muscle is presumably evenly loaded in life, and the force needed to tear it would give a poor indication of the force needed to break it. The average force which broke the insertion, in five experiments when a clean break occurred, was 9 kg wgt. The maximum lift that the muscle could resist without breaking its insertion is this force divided by the mechanical advantage of the muscle. This comes to about $0 \cdot 7$ kg wgt in gliding and $0 \cdot 5$ kg wgt in hovering. The mechanical advantages are different, in gliding and hovering, because of the different positions of the centre of lift.

The lifts that would be needed to break the insertion of the muscle are less than the lifts that would be needed to break the bones. The maximum lift that can be developed is probably smaller again, for the muscle is probably not strong enough to break its own insertion.

What are the largest lifts that are likely to act on a pigeon's wings? The pigeons used for the strength measurements weighed about $0 \cdot 4$ kg, so a lift of about $0 \cdot 2$ kg wgt on each wing would be needed for gliding in a straight line in still air (page 240). The lift generated by a hovering pigeon is probably never much more than the body weight, or about $0 \cdot 2$ kg wgt on each wing: if it were much more, hovering pigeons could accelerate rapidly upwards. The lifts in straight gliding and in hovering are apparently well within the limits allowed by the strength of the insertion of the pectoralis muscle. The same is probably true of straight flapping flight. The largest lifts likely to act on the wings are the lifts required to provide centripetal force in fast turns.

The insertion of the pectoralis major muscles would be broken in a glide by a lift of $0 \cdot 7$ kg wgt on each wing, or a total lift of $3 \cdot 5$ times the weight of the body. What sort of a turn would need this lift? The pigeon would have to be flying fairly fast, because the maximum lift a wing can give depends on the speed at which it is moving through the air (page 229). We will assume a speed of 15 m/s, and apply equation 37 (page 243) to find out how tight a corner the bird could turn, with a lift of $3 \cdot 5$ times its weight. The formula gives a radius of 7 m. I do not know how nearly the turning ability of pigeons approaches this limit, set by the strength of its muscle insertions, but I have yet to see a pigeon cripple itself by cornering too fast.

Balanced loads and bracing

We are going to see how the stresses in a structure can sometimes be reduced by making additional forces act on it. It will be convenient to regard tensile stresses as positive and compressive stresses as negative, since compression is negative tension.

Fig. 57a shows a pillar of radius r which is fixed at the base and has a crosspiece at the top. A downward force F acts on the right end of the crosspiece. This exerts a downward thrust F and a bending moment Fx on the pillar. The thrust, acting alone, would produce a stress $-F/\pi r^2$ in the pillar. The bending moment acting alone would produce stresses varying from $-Fxr/I = -4Fx/\pi r^3$ on the right face of the pillar to $+4Fx/\pi r^3$ on the left (the second moment of area for bending of a cylinder is $\pi r^4/4$, see Fig. 52). The total stress in the right face of the pillar is $-F/\pi r^2 - 4Fx/\pi r^3$ and must be negative (i.e. it must be a compressive stress). The total stress in the left face is $-F/\pi r^2 + 4Fx/\pi r^3$, and this will be positive (i.e. a tensile stress) if $x > r/4$.

Now suppose an additional downward force F acts on the left end of the crosspiece. If it acts at the same distance from the centre of the pillar there will be no bending moment, and there will be a uniform stress $-2F/\pi r^2$ in the pillar. Any tensile stress in the left face of the pillar will vanish and if $x > r/4$ the compressive stress in the right face will be reduced. (If the pillar were not solid but hollow with external and internal radii r_1, r_2, the condition would be $x > (r_1^2 + r_2^2)/4r_1$). A structure

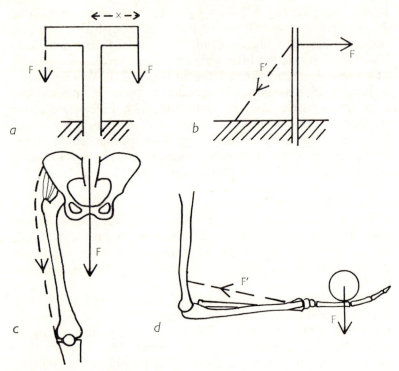

Figure 57. Diagrams illustrating the discussion of balanced loads and bracing

which would be broken by a single force F may yet be able to withstand two balanced forces F, acting together.

Fig. 57*b* shows a flag pole acted on by a horizontal force F, due to the wind. This bends the pole and if the wind gets too strong the pole will break. The chances of its breaking can be reduced by fixing a guy rope which will exert an oblique force F' on it, because this reduces the bending moment.

Pauwels (1948) drew attention to some applications in the human body of the principles illustrated in Fig. 57*a*, *b*. Consider a man who is standing on one leg or who is taking a step and has his weight supported by one leg. Muscles between the ilium and the dorsal end of the femur hold the femur in position. The weight of the body acts along a line some distance from the shaft of the femur and tends to bend it (Fig. 57c, where the force F represents the weight). The stresses in the femur are reduced by tension in the iliotibial tract, which is a strong fascia made of collagen fibres. It runs along the lateral face of the thigh from the ilium to the tibia, as indicated by the broken line in the figure, and has muscles to tighten it. It reduces the stresses in the femur in exactly the same way as the force on the second arm of the crosspiece in Fig. 57a reduces the stresses in the pillar.

Now consider a man holding a heavy weight, with his forearm horizontal (Fig. 57d). If the forearm were held only by the biceps and brachialis muscles, which insert on the radius and ulna near the elbow, large bending moments would act on the radius and ulna and the stresses in them would be large. The stresses will be reduced if the brachioradialis muscle contracts as well, because it inserts on the radius near the wrist and pulls in the direction indicated by the broken line in the figure. It has the same effect on the bones of the forearm as a guy rope has on a flagpole (Fig. 57b). The brachioradialis is used as well as the biceps and brachialis when the elbow is bent against a large force (Basmajian, 1962). The brachioradialis tends to bend the humerus, and it increases the stresses in the humerus at the same time as it reduces the stresses in the radius and ulna, so it would be disadvantageous from this point of view to have it as the sole flexor of the elbow. It would also be disadvantageous because a large muscle in the forearm would make the forearm unwieldy by increasing its moment of inertia about the elbow (see page 35).

Complex stresses

We distinguished in Chapter 3 (page 68) between tensile, compressive and shear stresses. Compression is negative tension, and tensile and compressive stresses are sometimes referred to together as direct stresses. It is usual to regard tensile stress as positive direct stress and compressive stress as negative direct stress.

Fig. 58a shows a square block of material. We shall see how various stresses deform it, and how they deform the part of the block which is bounded by a broken line. This is itself square, but its faces are at 45° to the faces of the block. In Fig. 58b the

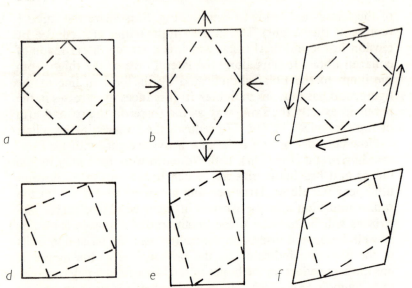

Figure 58. Diagrams showing that tensile strains involve shear, and *vice versa*

block is acted on by vertical tensile stress and horizontal compressive stress. These direct stresses have made it taller and narrower. The broken outline is deformed to a rhombus: in other words, it has been sheared. In Fig. 58*c* shearing stresses have been applied to the face of the main block, and the broken outline has been stretched so that it is rectangular. Direct stresses produce shearing stresses and shearing stresses produce direct stresses.

The broken outline in Fig. 58*a* is drawn with its sides at 45° to the faces of the main block. The one in Fig. 58*d* is drawn at a different angle. Direct stresses on the faces of the block do not make it a rhombus (Fig. 58*e*) and shearing stresses do not make it a rectangle (Fig. 58*f*), but both make it a parallelogram: both have the effect of shearing it and stretching it simultaneously.

All the stresses shown in Fig. 58 act in the plane of the paper and it will be convenient to restrict the scope of this paragraph to stresses in a single plane. For any point in a body subject to a two-dimensional system of stresses, two planes can be found at right angles to the plane of stress and to each other, in which no shearing stresses act. These two planes are called the principal planes for the point, and the direct stresses which act across them are the principal stresses. In other planes, there are shearing stresses as well as direct stresses. The shearing stresses are greatest in the planes at 45° to the principal planes, where their values equal half the difference between the principal stresses (in calculating the difference, tensile stresses should be taken as positive and compressive stresses as negative). When blocks of material are compressed, they often break along planes at 45° to the direction of compression: that is, along planes in which the shearing stresses have their maximum value.

When three dimensions are being considered, there are three principal planes mutually at right angles and three principal stresses.

In the tests on pigeon bones described in the preceding section of this chapter, bending moments and twisting moments were applied separately. How much greater would the stresses have been, if they had been applied together? There are equations given in engineering textbooks (such as Warnock and Benham, 1965) which can be used to solve problems like this. Using them I estimate that if the bending moment which broke the humerus and the twisting moment which broke the humerus had been applied together, the maximum direct stress would have been 12% higher than when the bending moment alone was applied. This does not enable us to calculate a combination of bending and twisting moments that would just break the bone, because twisting would have altered the direction of the maximum stress as well as its size, and bone has different strengths in different directions.

Photoelasticity—isoclinics

Photoelasticity is a property of transparent high polymers which can be used to get information about stresses that cannot be calculated. Engineers have found it particularly useful in studies of stress concentration.

Ordinary light can be thought of as involving vibrations in all directions at right angles to its path. Any of these vibrations can be resolved into components in two directions at right angles to each other. Some materials, which are described as birefringent, resolve light passing through them into two components, each involving vibration in one direction only, which travel at different speeds. The directions of vibration in the two components are at right angles.

A sheet of polaroid is transparent only to the component of light whose vibrations are in one direction, known as the optic axis of the polaroid. Light which has passed through a sheet of polaroid involves vibrations in only one direction and is called plane polarized light. If two sheets of polaroid are put one behind the other with their optic axes at right angles, no light can get through because the component of the light which gets through the first sheet is precisely the component that is blocked by the second. Two sheets of polaroid arranged in this way are said to be crossed.

Light can pass through a pair of crossed polaroids, when there is a birefringent material between them. This is illustrated by Fig. 59. Each of the diagrams shows the direction or directions of vibration in a ray of light, seen end on. Suppose the light passes first through a polaroid whose optic axis is vertical. The light that gets through will involve vertical vibrations only (Fig. 59a). If it then passes through a birefringent material, it will be resolved into two components at right angles (Fig. 59b). Each of these components has itself got a horizontal component (Fig. 59c). These horizontal components can pass through a second polaroid whose optic axis is horizontal. If they were in phase they would cancel each other out and no light would pass

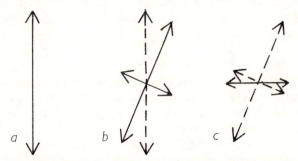

Figure 59. Diagrams showing how light can pass through crossed polaroids if they have birefringent material between them. These diagrams are explained in the text

through this polaroid, but they will generally not be in phase because the two components represented in Fig. 59*b* travel through the birefringent material at different speeds.

One can find out whether a material is birefringent by putting it between crossed polaroids or Nicol prisms (which have the same effect on light) and looking to see whether light can pass through. If the specimen is small, this is best done in a polarizing microscope. One can also find out the directions of vibration of the components into which it splits the light, by rotating the specimen. Although light can pass through in most positions, it will not do so when the directions of these components exactly match the directions of the optic axes of the polaroids. Look at Fig. 59 again. If the birefringent material were arranged so that it split light into components whose vibrations were vertical and horizontal, the vertically polarized light from the first polaroid would travel through it unchanged and there would be no horizontal components to pass through the second polaroid.

Birefringence depends on the arrangement of molecules. Transparent amorphous polymers are not birefringent when they are not stressed, because their molecules run at random in all directions. When they are stressed the molecules tend to line up along the direction of the principal tensile stress. They become birefringent, splitting light into components with vibrations in the direction of the principal tensile stress and at right angles to it (i.e. in the two principal planes). This is the phenomenon of photoelasticity. It can be seen, for instance, by stretching resilin under a polarizing microscope. Though it is not initially birefringent, it becomes birefringent when it is stretched (Weis-Fogh, 1960). Fibres have their molecules more or less aligned even when they are not stressed and they are birefringent even when they are not stressed.

Photoelasticity can be used to discover the directions of stresses. If the structure being investigated was made of a transparent amorphous polymer, it might be possible to investigate it directly. Normally one has to make a model of the structure from a material such as Perspex or Araldite, and investigate stresses in this model. The technique is illustrated in Fig. 60. A is a light source giving a wide parallel beam. BB are a pair of polaroids which can be rotated about the axis of the

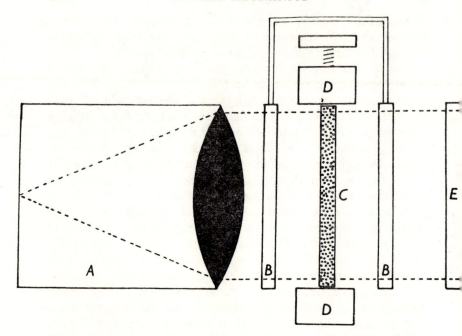

Figure 60. Apparatus for examining isoclinics in a stressed transparent model. The labels are explained in the text. (From Smith, 1962a)

light beam but are fixed together so that they always remain crossed. C is the model and D is a clamp which can be used to apply stresses to it. E is a ground glass screen. When the model is stressed it becomes birefringent and light passes through all parts of it to the screen, except the parts where the principal stresses are parallel to the optic axes of the polaroids. These parts will appear on the screen as dark lines, known as isoclinics, on an otherwise bright image of the model. If the polaroids are rotated by stages through 90° a series of sets of isoclinics will be seen and the principal planes of all parts of the model can be found. The technique does not distinguish between tensile and compressive stresses but merely gives a pair of directions at right angles for each part of the model. A little thought is usually enough to decide which directions represent tensile stresses and which compressive ones.

The technique described can only be applied to two-dimensional models. It is possible to obtain isoclinics for three-dimensional models by a more elaborate technique (see, for instance, Warnock and Benham, 1965).

Epiphysial plates and bony trabeculae

A limb bone from an immature mammal is not a single chunk of bone, but has separate pieces at its ends attached only by thin layers of cartilage. The separate pieces of bone are known as epiphyses, and the layers of cartilage as epiphysial plates. The articular cartilages of the joints are not supported directly by the main body of the bone but by epiphyses. Many tendons insert on epiphyses (Barnett and Lewis, 1958). The arrangement is thought to make growth interfere less with joints and insertions, than it would if the bone grew as one piece.

Cartilage is weaker than bone, so the epiphysial plates are weak regions in a young bone. There might be a danger that stresses would break epiphyses off. Compressive stresses acting across the plates would not do this; rather, they would press the epiphyses firmly into place. Tensile stresses and shear stresses might be dangerous. Smith (1962a, b) studied isoclinics in Perspex models of limb bones to find the directions of stresses likely to occur in the epiphysial plates.

Fig. 61a shows the model he used to study the stresses in the human heel. There is an epiphysis on the back of the heel, on which the Achilles tendon inserts. The model is cut from a sheet of Perspex and its main outline is that of a saggittal section of the tarsals and metatarsals. There is a hinge at E, representing the joint between the tarsals and metatarsals. The projection D represents the Achilles tendon and the projections C and H, joined by a cord, represent the ligaments and muscles of the sole of the foot. The model is shown arranged to simulate the stresses which occur in running and at the stage of a walking step when the heel is off the ground. A downward and somewhat backward force A is applied at the position of the ankle joint, and an upward pull at D simulates tension in the Achilles tendon. These forces automatically tighten the cord CH since the anterior end of the model rests on a rigid support.

Fig. 61b shows the heel with nine sets of isoclinics. The corresponding directions of the optic axes of the two polaroids (the

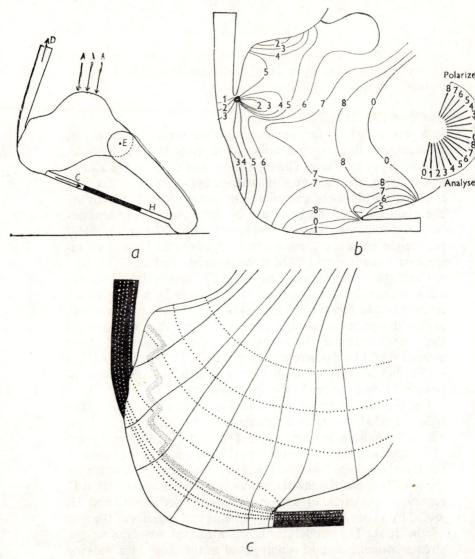

Figure 61. The results of an investigation of the stresses in the human heel, using the apparatus of Fig. 60. (*a*) The model, in the position in which it was used with the forces which were applied to it indicated by arrows. (*b*) The isoclinics which appeared in the heel when the optic axes of the polaroids were at the angles indicated by the numbers. (*c*) The pattern of stresses deduced from the isoclinics. Continuous lines show the directions of principal compressive stresses, and dotted lines the directions of principal tensile stresses. Black areas represent soft tissues and the stippled band represents the epiphysial plate. (From Smith, 1962a)

polarizer and analyser) are indicated. Fig. 61c shows the directions of the principal stresses, deduced from the isoclinics. The stresses running along the dotted lines must be tensile stresses and those running along the continuous lines must be compressive stresses. Tensile stresses act along the Achilles tendon and the ligaments of the sole, and round the heel between them. Compressive stresses act radially in the heel, everywhere at right angles to the tensile stresses.

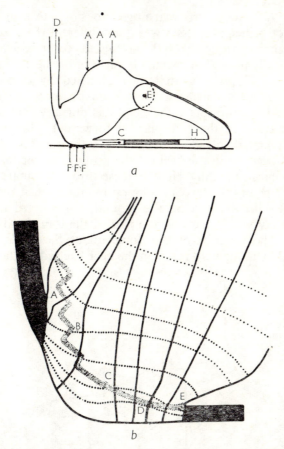

Figure 62. An analysis of the stresses in the heel in standing with the heel on the ground. (*a*) The position of the model and the forces applied to it. (*b*) The pattern of stresses. Other details as in Fig. 61. (From Smith, 1962a)

The position of the epiphysial plate is shown by stippling in Fig. 61c. Anteriorly it runs parallel to the tensile stresses and perpendicular to the compressive ones. The compressive stresses acting across it will tend to hold the epiphysis firmly in place. Posteriorly it forms a series of steps. It runs alternately parallel to the tensile stresses and to the compressive ones, so that it is nowhere parallel to shearing stresses. This is probably the best possible arrangement, to minimize the danger of the epiphysis being broken off.

When the model was rearranged to simulate standing the pattern of principal stresses was quite different (Fig. 62). Since the heel presses vertically on the ground the compressive stresses run vertically instead of radially in most of the heel. The line of the epiphysial plate is no longer matched by directions of principal stresses. Shearing stresses act on most parts of the plate, tending to displace the epiphysis dorsally (but at E the shearing stress tends to displace it forwards). The epiphysial plate is adapted to reduce danger from the stresses which are likely to occur with the heel off the ground, rather than those which occur in standing. How do these stresses compare in size?

Fig. 63 shows the forces acting on my foot, when I stand on one foot with the heel off the ground. An upward force W, equal to my weight, acts on the part of the foot that is on the ground. The moment of this force about the ankle joint is balanced by an upward force on the heel, exerted by the gastro-

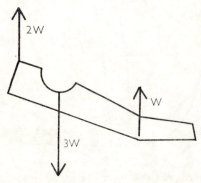

Figure 63. A diagram showing the forces that act on the foot, when a man stands on one foot with the heel off the ground

cnemius muscle through the Achilles tendon. The mechanical advantage of the muscle is $0\cdot5$, so the force must be 2 W. The downward force on the ankle joint must be 3 W to balance the upward forces. Much larger forces could act in running and jumping. If I jump down from a height of 3 ft and land on one foot I will decelerate rapidly and probably exert a force of at least 3 W on the ground, so the force at the ankle joint will be 9 W. These forces are much more than the forces which act in standing. Even if I stand on one foot the force at the ankle joint will only be W or a little more, depending on whether my centre of gravity is directly over the joint or not. Much larger stresses are likely with the heel off the ground, than with the heel on the ground. The epiphysial plate is adapted to reduce danger from the pattern of stresses which is likely to involve the largest stresses.

Smith found similar arrangements at other epiphyses. They seemed to be arranged as far as possible so that compressive stresses acted at right angles to the epiphysial plates. This is not possible at the top of the tibia, where the patellar tendon inserts on an epiphysis. When the tendon is taut tensile stresses act at right angles to the epiphysial plate, tending to tear the epiphysis off. The cartilage of the epiphysial plate is appropriately replaced here by bundles of collagen fibres running from the body of the bone to the epiphysis in the direction of the tensile stresses, and holding the epiphysis in place (Smith, 1962b).

Many bones have regions of spongy bone, where the bone forms only a network of fine trabeculae among soft tissue. The human heel bone is an example. The trabeculae do not run in all directions at random, but tend to form lines running in the directions of the principal compressive stresses shown in Fig. 61. Long slender struts are apt to buckle when compressive stresses act along them (page 156) but the trabeculae have cross-connections which must tend to prevent them from buckling. The trabeculae run in the directions of the large compressive stresses which occur with the heel off the ground, rather than of the smaller ones which occur in standing (Smith, 1962a). In the tibia, under the insertion of the patellar tendon, the main trabeculae run along the lines of principal tensile stress. These trabeculae have relatively few cross-connections and would presumably be weak in compression (Smith, 1962b).

6—AM

Photoelasticity—isochromatics

Light can only pass through crossed polaroids with bire-fringent material between, because the material splits the light into two components which travel through it at different speeds and emerge out of phase (page 142). If it so happens that the slower component is delayed relative to the faster one by a whole number of wavelengths, they will emerge in phase and no light will get through the second polaroid. The amount of delay depends on the thickness of the material and on the degree of birefringence. In materials made birefringent by stress, it is proportional to the difference between the two principal stresses. This phenomenon is exploited in a technique for obtaining quantitative information about stresses.

The technique uses models like those used for finding the directions of stresses by isoclinics. However, the apparatus is modified so that isoclinics are no longer formed (this is ex-plained in textbooks such as Warnock and Benham, 1965). If sufficiently large stresses act in the model to delay the slow component of the light by one or more wavelengths, the model looks stripy. It is dark wherever the delay is a whole number of wavelengths, and the dark stripes connect points which have equal differences between the principal stresses. They are known as isochromatic fringes. One can decide how many wavelengths delay have produced each fringe in a complex pattern, if one starts with small forces acting on the model and increases them by stages, observing the fringes at each stage, until the complex pattern is obtained. The information obtained directly from isochromatic fringe patterns is the difference between the principal stresses at each point in the model. Determination of the stresses themselves, rather than their difference, sometimes involves quite elaborate analysis (see, for instance, Warnock and Benham, 1965).

Pauwels (1948) used isochromatic fringe patterns to illustrate the way the iliotibial tact reduces stresses in the femur, and the way the brachioradialis muscle reduces stresses in the bones of the forearm (page 138). Fig. 64a shows a fringe pattern obtained in a model of the femur, in which the iliotibial tract was not represented. The force which represents the body weight acted along the broken line, and bent the femur. It set up longitudinal

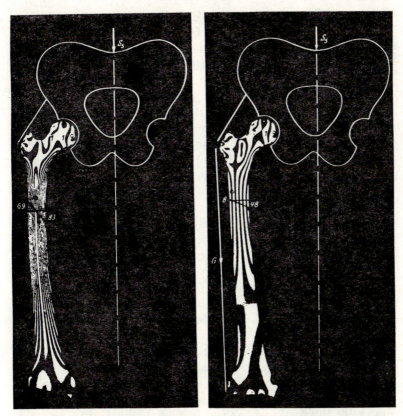

Figure 64. Isochromatic fringes in a model of the human femur loaded to simulate standing (*a*) without, and (*b*) with a brace representing the iliotibial tract. (Traced from photographs in Pauwels, 1948)

direct stresses in the shaft of the femur but no transverse ones, so the isochromatic fringes are contour lines of longitudinal stress and not merely of difference between the principal stresses. The fine dark line in the middle of the femur, indicated by a small circle, marks the neutral surface where no stresses act. The dark fringes on either side of it mark successive increments of 10 kg/cm^2 of stress. The numbers show the stresses at the edges of the femur at one particular level, obtained by counting the fringes. Note that the stress in the outer face of the femur is a tensile stress. Fig. 64*b* shows a fringe pattern in the same model under the same load, after a brace representing the iliotibial tract had been fitted. The tensile stress in the lateral edge of the femur almost vanished and the compressive stress in the medial edge is greatly reduced.

The fringe patterns do not provide reliable estimates of the stresses that would occur in a real femur, because the two-dimensional model required by the technique is so unlike a real femur. Much better estimates could be obtained by mathematical analysis. The value of the fringe patterns in this particular investigation was that they provided striking visual demonstrations of the engineering principles involved.

Measurement of surface strain

Photoelastic investigations provide information about stresses and strains in models of structures. There are several techniques for investigating strains at the surface of the structure itself. The simplest uses a grid drawn on the structure, or some other pattern of surface markings. The structure is photographed while it is stressed and while it is unstressed, and the photographs are compared. Changes in the distances between markings show the strains at various parts of the surface. Another technique involves painting the structure with a brittle lacquer such as "Stresscoat". When the structure is stressed, cracks appear in the lacquer at right angles to the direction of principal tensile strain. If the stress is applied gradually, the cracks will appear first where the strain is greatest. Another technique involves resistance strain gauges.

Stretching alters the resistance of metal wire. Strain gauges are made of very thin wire or metal foil bonded on to paper. They are cemented on to the surface of the specimen. When the specimen is stressed the gauges are strained and their resistances change. The changes are measured on a Wheatstone bridge. A single strain gauge is sensitive to strain along a single axis, but the size and directions of the two principal strains at a point on the surface of a specimen can be determined by means of a rosette gauge, which consists of three strain gauges set at angles of 45° or 60° to each other (Warnock and Benham, 1965).

Biting strains

Endo (1965) used resistance strain gauges to investigate the strains that occur in the human skull when something is bitten. It was naturally impossible to attach strain gauges to a living person's skull. He used dry skulls with strain gauges cemented in various positions.

The jaws of man, like those of other mammals (page 5) are closed by temporalis, masseter and pterygoideus muscles. Fig. 65 shows how Endo simulated the forces exerted by the temporalis and masseter muscles. (He ignored the pterygoideus muscles.) Pieces of canvas (shown hatched) were glued to the skull where the muscles originated (shown stippled). The fossae where the lower jaw articulates were placed on firm supports and the free ends of the two canvas sheets on each side of the skull were attached to the two ends of a lever which was anchored in the manner indicated. A lever was used to apply an upward force Pd to a single tooth, as if that tooth alone of the upper teeth was being used to make a bite. This force was automatically balanced by forces Pa at the mandibular fossae and Rt, Rm in the pieces of canvas representing the temporalis and masseter muscles. The manner of anchoring the pieces of canvas ensures that the forces Rt, Rm act in the proper directions. The pattern of surface strain in a skull stressed in this way is probably reasonably natural.

Two examples of Endo's results are shown in Fig. 66. A force applied to the first molar (Fig. 66, left) produces a different pattern of strain from a force applied to the first incisor (Fig. 66, right). The strains on the same side of the skull as the loaded tooth are different from the strains on the other side. There are relatively large strains in the upper jaw, in the zygomatic arch and around the eye, but the strains in the forehead are small. The bones of the forehead of course have to protect the brain and are probably much thicker than would be necessary if they served simply to transmit the forces involved in biting.

The masseter pulls downward on the zygomatic arch, so there are roughly vertical tensile strains at its origin. The downward pull of the masseter and the upward force P on the tooth act together to produce shearing stresses under the eyes, where the principal stresses and strains consequently run at about 45° to

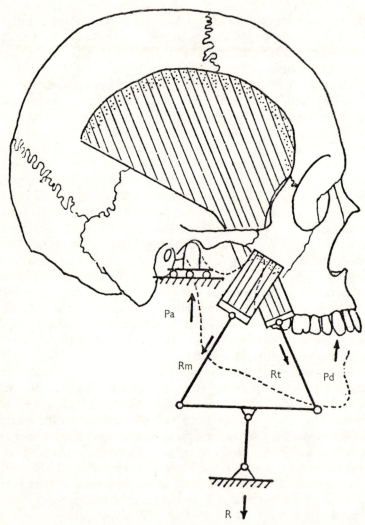

Figure 65. Apparatus used for simulating, in a human skull, the stresses
which occur when food is bitten. (From Endo, 1965)

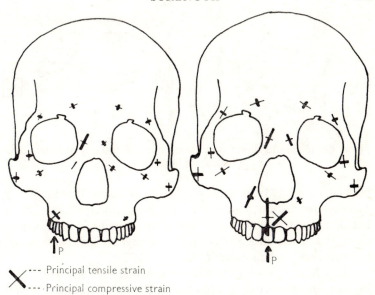

--- Principal tensile strain

---- Principal compressive strain

The lengths of the lines indicate the amounts of strain

Figure 66. Strains at the surface of a human skull in the apparatus of Fig. 65 with the force *P* (4·5 kg) applied to two different teeth. (Redrawn from Endo, 1965)

the vertical. These stresses would be greater, were it not for the lateral walls of the orbits. These are in tension, and apparently act as stays for the zygomatic arch as well as protecting the eyes.

Elastic stability

Equilibrium can be stable or unstable. Fig. 67*a* shows a bar hanging from a pivot. Its weight is balanced by the reaction at the pivot, and it is in equilibrium. If the rod is pushed to one side, as indicated by the broken outline, the equilibrium is disturbed. The weight of the bar comes to have a moment about the pivot, tending to swing it back to the equilibrium position. This sort of equilibrium is described as stable. Fig. 67*b* shows the same arrangement inverted. When the bar is exactly vertical its weight and the reaction of the pivot are in line, and it is in equilibrium. If it is pushed even very slightly to one side, the weight has a moment about the pivot which tends to swing the

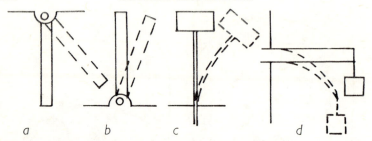

Figure 67. Diagrams illustrating the meaning of stability and the account
of elastic instability

bar further away from the equilibrium position. If the rod is put
in the equilibrium position it will not remain there, because
accidental disturbances are bound to occur (it is assumed that
there is not too much friction at the pivot). This sort of equilibrium is described as unstable.

Fig. 67c shows a long slender rod which is fixed rigidly at the
bottom and has a weight attached at the top. The rod is flexible.
If it is bent to one side, as indicated by the broken outline, the
weight exerts a bending moment tending to bend the rod more,
while the elastic restoring forces in the rod tend to straighten it.
If the weight is less than a certain amount, the restoring forces
will overcome the bending moment for any small deflection. The
rod will straighten again after it has been bent and the equilibrium with the rod straight will be stable. If the weight is more
than this critical amount, the bending moment will overcome
the restoring forces, for any small deflection. The equilibrium
with the rod straight will be unstable, and the rod will always
bend to the side. It can be shown that if the length, l, of the rod
is more than about 30 times its diameter the critical downward
force is $\pi^2 EI/4l^2$, where E is Young's modulus and I is the
second moment of area of the cross-section about the neutral
axis for bending. If the rod is not round, the value of I that
should be used is the value for the direction in which it is most
easily bent. We will use this expression when we discuss insect
plastrons (page 203). It is only applicable, if the force acts
directly along the axis of the rod when the rod is straight. If the
force acts to one side of the axis so that there is a bending
moment even when the rod is straight, a more complicated

formula must be used which involves the strength of the material as well as its modulus. The expression given here only applies to situations exactly like that represented in Fig. 67c, with one end of the rod rigidly fixed and the other entirely free. Expressions relating to slender struts with their ends fixed in other ways are given in engineering textbooks (for instance, Warnock and Benham, 1965).

The type of instability we have been discussing is called elastic instability. Because of it, the maximum load which a slender column will support is far less than the load which would be needed to give compressive stresses equal to its compressive strength. The load a short fat column will support is limited by its compressive strength, but the load a tall thin one will support is limited by elastic instability.

The phenomenon of elastic instability is by no means limited to slender columns (Timoshenko, 1936). If a hacksaw blade is clamped in a vice so as to project horizontally with its teeth pointing vertically, loads up to a certain weight can be hung from the free end without bending it much. If the load exceeds this weight, the blade twists to one side and bends down, owing to elastic instability (Fig. 67d). If a heavy load is set on the end of a vertical, thin walled tube, the tube may become unstable and crumple like a concertina. If a thin walled tube is bent, it may become unstable and crumple on the compressed side.

Concertina-like folds are only possible in tubes with very thin walls. A different form of elastic instability can occur in tubes with thicker walls. It can be demonstrated on the rubber tubing used for bunsen burners and (rather better) on plastic drinking straws. When the tube is bent, the wall on the outer side of the bend is stretched and the wall on the inner side compressed. Because of this the cross-section of the tube does not remain circular, but becomes flattened into an oval. The bending moment is of course resisted by an elastic restoring moment. In the early stages of bending, the restoring moment increases as bending proceeds. This does not continue indefinitely, because of the flattening of the cross-section. A point is reached beyond which further bending does not increase the restoring moment, but decreases it. At this point the tube becomes unstable and collapses suddenly with a kink in it. Brazier (1927) has derived an equation for the critical bending moment, and an approximate form of his equation is given below. A tube of radius r (measured to a point half-way through its wall), with a wall of thickness t made of a material of Young's modulus E, will become unstable at a bending moment M where

$$M \simeq 1 \cdot 1 \ Ert^2 \qquad (16)$$

We have already enquired how strength in bending can best be combined with lightness (page 128). We saw that a beam of given length which is to carry a given load, can be made lighter if it is made hollow than if it is made solid. A wide tube with a thin wall can be made lighter than a narrower tube which would need a much thicker wall. This only applies when strength in bending is limited by the tensile or compressive strength of the material. It does not apply to tubes which fail by elastic instability. Consider the case of tubes that fail by kinking to which equation (16) applies. Since the cross-sectional area, of the wall, A, is $2\pi rt$ we can re-write the equation

$$M \simeq 1 \cdot 1 \ E(rt)^{3/2}(t/r)^{1/2} = 1 \cdot 1 \ E(A/2\pi)^{3/2}(t/r)^{1/2}$$

This gives

$$A^3 \simeq 8 \ \pi^3 M^2 r/1 \cdot 2 \ E^2 t$$

and A decreases as t/r increases if M is kept constant. A fairly narrow, fairly thick-walled tube can be made lighter than a wider, thinner-walled one, to withstand the same bending moment.

It seems that there are optimum proportions for a tube which is to be as light as possible and yet withstand a given bending moment. Up to a point it is advantageous to make it wider and thinner-walled. Beyond that point it is disadvantageous. The critical point at which the proportions of the tube are ideal, is the point at which it changes from failing by breaking, to failing by elastic instability. The ideal proportions will depend on the ratio of Young's modulus (which affects elastic stability) to the tensile strength. It can be estimated from the information given here and on page 129 that if the modulus were 100 times the tensile strength the tube would be best made with the thickness of its wall about $0 \cdot 03$ of the radius, but if the modulus were 50 times the tensile strength the thickness of the wall should be about 0·06 of the radius. Bone and insect cuticle both have Young's modulus about 100 times the tensile strength, but it would be rash to conclude that bones and insect legs should have radii 30 times the thickness of the wall because bone and insect cuticle probably have different moduli in different directions and because no account was taken on page 129 of the effect on strength of the flattening of the cross-section which occurs in bending.

Currey (1967) has done a calculation similar to this, based on an equation which is not quite the same as equation (16).

Elastic stability of the locust tibia

The greatest forces which the hind legs of locusts have to withstand are probably the forces involved in making a strong jump. We have already estimated these forces (page 25). We shall now see how the tibia is suited to withstand them. Its shape and dimensions have probably been determined almost entirely by natural selection related to its mechanical function. There is certainly no question of its shape being determined by its contents, for it contains little but blood. In this it is quite different from the femur which is swollen to accommodate the muscles of the "knee" joint.

The forces which act on the tibia at the start of a jump tend to bend it. It is a tube of hard cuticle and if it is bent too far it gives way in the same way as a drinking straw, forming a single kink. The maximum bending moment it can withstand is not determined by the strength of the cuticle, but by elastic instability.

The maximum bending moment can be determined very easily. The tibia is cut from the hind leg of a freshly killed locust and its proximal end is pushed into a metal tube filled with dental cement. When the cement has set the tube is clamped in a retort stand so that the tibia is horizontal with its posterior side up. Weights hung from its free end will then bend it in the same direction as the forces which act in jumping. A light pan is hung from the end of the tibia and weights are added gently until the tibia gives way. My students and I have done this with a number of locust tibias, and have got very varied results. The tibias of very young adults often give way under a load of five times the body weight but those of older adults may withstand loads of 50 or more times the body weight. This is because the cuticle of the tibia gets thicker with age (Jensen and Weis-Fogh, 1962).

The loads in these experiments were applied close to the end of the tibia and at right angles to it. In jumping, the line of action of the reaction of the ground passes close to the end of the tibia, but is only at right angles to it in the early stages of the jump before the full force has been developed (Brown, 1963a). By the time the position shown in Fig. 12b has reached the reaction must be about 8 times the body weight but it is no longer at right angles to the tibia and its component at right angles to the tibia is only about 6 times the body weight. This is well within the range of loads that kinked the tibias of young adults, but well below the loads needed to kink the tibias of older ones. Young adults can jump about as well as older ones. The results of the crude experiments suggest that the tibias of young adults are barely able to withstand the forces involved in jumping. However, the loads were left hanging from the tibias for some seconds in the experiments, and the tibias might have stood greater loads applied momentarily.

The most proximal parts of the tibia would suffer the biggest bending moments in a jump. They were embedded in dental cement in the experiments, and the kink formed just outside the cement. This was about $1 \cdot 5$ cm from the distal end of the tibia so the bending moment with a load of L g would be $1 \cdot 5$ Lg dyn cm. The tibia is not a strictly cylindrical tube, but it is sufficiently nearly cylindrical for it to seem reasonable to apply equation (16) to it. Its radius is about $0 \cdot 05$ cm. Pieces of its wall, taken from the tibias of older adults near where they kinked, were measured with a micrometer gauge and found to be about $0 \cdot 005$ cm thick. Young's modulus for the cuticle of locust tibia, stretched lengthwise, is about 10^{11} dyn/cm^2 (Jensen and Weis-Fogh, 1962). This is not the modulus we really want, for kinking involves a change from a roughly circular cross-section to a roughly elliptical one, which will depend on the modulus for circumferential stretching, which may be different. We will use the value of 10^{11} dyn/cm^2 for lack of further information. Equation (16) then gives

$$1 \cdot 5\,Lg \simeq 1 \cdot 1 \times 10^{11} \times 5 \times 10^{-2}\,(5 \times 10^{-3})^2$$
$$L \simeq 90\ \text{g}$$

This is the load we would expect on theoretical grounds to make the tibia kink. It lies within the range of critical loads found in the experiments.

The diameter of the tibia is nearly constant along its length but its wall is about twice as thick at the proximal end as at the distal end (Jensen and Weis-Fogh, 1962). This is appropriate, for it is at the proximal end that the bending moments are highest. According to equation (16), doubling the thickness of the wall should increase the critical bending moment by a factor of 4.

Hardness

The term "hardness" is used for several different but related properties which are measured in different ways. One of these properties is the ability to scratch other materials. Diamond will scratch glass but glass will not scratch diamond, so diamond is said to be harder than glass. In general, if material A scratches material B, material B will not scratch material A. Moh's scale of hardness is a list of materials in increasing order of hardness. Each material will scratch all the materials given lower numbers in the scale, and is scratched by all the ones with higher numbers. It is

1	talc	6	orthoclase
2	gypsum	7	quartz
3	calcite	8	topaz
4	fluorite	9	corundum
5	apatite	10	diamond

The hardness of a material can be gauged by finding out which of these materials will scratch it, and so finding its position on the scale.

Another property called scratch hardness is measured by moving a weighted diamond point across the surface of the specimen. The width of the scratch it makes is measured and the load on the point is divided by this width to obtain the hardness (see Ritchie, 1965).

A third property called indentation hardness is measured by

pressing a hardened steel ball or a diamond point against the specimen. The hardness is obtained by dividing the load on the ball or point by the area or depth of the indentation it makes. The Brinell, Vickers and Rockwell hardness tests all work in this way, but differences between the types of indenter and procedures used have made it difficult to devise a satisfactory means of converting values obtained in one of these tests, to values which would be obtained in the others (Warnock and Benham, 1965). Special difficulties are involved in measuring indentation hardness of visco-elastic materials (Ritchie, 1965).

Radulae of browsing molluscs

The chitons and many shore-living gastropods feed on fila-mentous algae which grow on the surfaces of rocks. They get the algae off the rocks by means of the radula, which is an organ covered by tiny teeth. It can be protruded from the mouth and is used like a rasp (Märkel, 1964).

The algae form a thin layer on the rock surface, adhering to it closely, growing down into hollows in it and even boring into it to a depth of a few millimetres (Lowenstam, 1962b). They bore particularly in limestone. Molluscs whose radular teeth are harder than the rock, in the sense of being higher on Moh's scale, can scrape the rock away and get at boring algae and algae in hollows. Molluscs whose teeth are softer than the rock can only scrape off the superficial layer of algae.

Some alga-eating gastropods such as the winkles (*Littorina*) have radular teeth made entirely of transparent organic material (Lowenstam, 1962b). They scratch gypsum but not calcite, so their hardness is between 2 and 3 on Moh's scale. They are too soft to scratch most rocks, including limestone (which consists mainly of calcite). Mineral particles may be found in winkle gut contents, but they are presumably particles which were already loose and not ones scraped off by the radula. Winkles can presumably only feed on the superficial layer of algae.

Some other gastropods including limpets (*Patella*) have radular teeth capped by opaque material shown by X-ray diffraction to be the mineral goethite, $Fe_2O_3.H_2O$. The hardness of this mineral is 5–5·5 on Moh's scale (Hodgman, 1965) and the hardness of the teeth is apparently in this range, for they will just scratch apatite. Molluscs with teeth as hard as this can scrape away limestone and obtain burrowing algae.

The radular teeth of chitons are capped by magnetite, Fe_3O_4, whose hardness is 6 on Moh's scale (Lowenstam, 1962a). They can of course scratch limestone but they are still softer than many rocks. Chitons living on hard rocks like granite often have well worn radular teeth. The rock is not scratched by them but, on the contrary, wears them down. Chitons living on limestone often have no signs of wear on their radular teeth, and grooves scraped by the teeth can be found on limestone rocks where chitons live.

CHAPTER 5

PRESSURE, DENSITY AND SURFACE TENSION

THE physical topics dealt with in this chapter are all branches of hydrostatics, but some of them are discussed in textbooks on physical chemistry (for instance, Glasstone and Lewis, 1960) as well as in books on physics.

Pressure

The pressure at a given position in a fluid is the force which would act on unit area on either side of a plane surface in that position. It is the same in all directions, so long as the viscosity of the fluid can be ignored (Ramsey, 1946); it is exactly the same in all directions for stationary fluids and for most practical purposes it can be taken as the same in all directions in moving fluids. Since pressure is force per unit area it can be expressed in dyn/cm^2 or μ bar but it is sometimes convenient to express large pressures in terms of atmospheric pressure at sea level, which is about 10^6 dyn/cm^2 (1 bar). A pressure of P dyn/cm^2 can therefore be described as a pressure of $10^{-6} P$ atm (strictly, as $0\cdot988 \times 10^{-6} P$ atm; see Nelkon and Parker, 1965).

There are many methods of measuring pressure. One of the most straightforward is with a manometer, which is a U-tube partly filled with liquid. If one arm of a manometer is connected to a region where the pressure is P_1 and the other to a region where the pressure is P_2, the manometer settles with the levels of liquid in its arms differing by a height h, such that

$$P_1 - P_2 = \rho g h \qquad (17)$$

where ρ is the density of the liquid. The liquids most often used in manometers are water and mercury, so it is often convenient to describe a pressure as h cm water or mercury. A pressure of 1 cm water is very nearly 10^3 dyn/cm^2 and a pressure of 1 cm mercury is $1\cdot33 \times 10^4$ dyn/cm^2.

For some purposes, manometers are unsatisfactory. They will not work properly if the U-tube is too fine, because of surface tension effects (the difficulty is due to the difference between advancing and retreating contact angles, see page 200). Hence, an appreciable mass of liquid must be used and its inertia may prevent the manometer from following rapid changes of pressure. Also, a manometer will affect the pressure it is supposed to be measuring, if this is the pressure in a smaller container. For instance, if a manometer was used to measure the pressure in a small worm and the worm contracted its muscles so as to increase the pressure, some of its contents would be squeezed out to work the manometer, and this would tend to reduce the pressure again. In general, manometers are only useful for measuring constant or slowly changing pressures in fairly large containers.

Pressure transducers can be bought, which can be used to measure more rapidly changing pressures or pressures in smaller containers. They contain small metal diaphragms. The pressure which is to be measured is applied to one side of the diaphragm and distorts it slightly. The distortion is measured electrically, for instance by measuring the electrical capacity between the diaphragm and a metal plate just behind it. Pressure transducers can be made to provide a continuous record of pressure changes, by connecting them to a cathode-ray oscilloscope, a pen recorder or an ultra-violet recorder.

If the contents of a container are at a different pressure from the containers' surroundings, forces will act on the walls of the container. Consider a boiler with a domed top. If the pressure in the boiler is P above atmospheric pressure, what is the force tending to blow the dome off? A force P per unit area acts everywhere on the curved surface of the dome, but it acts in different directions at different places because it always acts at right angles to the surface. The resultant upward force on the dome is not P times the area of the curved surface of the dome, but P times the area of a full-scale plan of the dome. More generally and in more mathematical terms, the thrust in a given direction on a curved surface is the product of the pressure and the area of the surface projected on a plane at right angles to the direction in question.

Now consider the forces acting in the walls of a tube of radius r, when the pressure in the tube is greater than the pressure outside the tube by P. Think of the tube as being split lengthwise down the middle. What force will there be to push the two halves apart? If the length of the tube is l the projected area of each half is $2rl$ and the force will be $2rlP$. This can be expressed as a tension: that is as a force per unit length. The total length of the cut edges of each half of the tube is $2l$ so the tension T is $2rlP/2l$ and

$$P = T/r \qquad (18)$$

T is the tension acting round the circumference of the tube. If the tube has ends there will be forces $\pi r^2 P$ tending to blow them off. These will cause lengthwise tension in the wall of the tube. As the circumference of the tube is $2\pi r$ the lengthwise tension T' will be $\pi r^2 P/2\pi r$ and

$$P = 2T'/r \qquad (19)$$

The tension in the wall of a sphere of radius r inflated by a pressure P is given by the equation

$$P = 2T/r \qquad (20)$$

There is also a simple formula for the tension in container walls which curve with unequal radii in two directions (Cottrell, 1964).

Hydrostatic skeletons

The principal of hydrostatic skeletons is very simple and it is very important, especially in the movement of soft-bodied animals (Chapman, 1958; Clark, 1964). Consider a hollow cylindrical animal filled with liquid. Suppose it has longitudinal muscles in its body wall running parallel to the axis of the cylinder and circular muscles running round its circumference. If no liquid is allowed to enter the body cavity or to escape from it the volume of the animal will be virtually constant, for liquids are practically incompressible. If the circular muscles contract the animal must get longer as well as thinner, and if the longitudinal muscles contract it must get fatter as well as shorter. Hence, the circular and longitudinal muscles are antagonistic to each other. If the longitudinal muscles of one side contract and the circular muscles are used to prevent the animal from getting fatter, the animal will bend to that side and the longitudinal muscles of the other side will be stretched.

A closed, liquid-filled body cavity makes it possible for muscles to act antagonistically to each other. Since a skeleton of rigid levers such as bones makes it possible for muscles to act antagonistically as flexors and extensors, the term "hydrostatic skeleton" has been coined for body cavities which promote

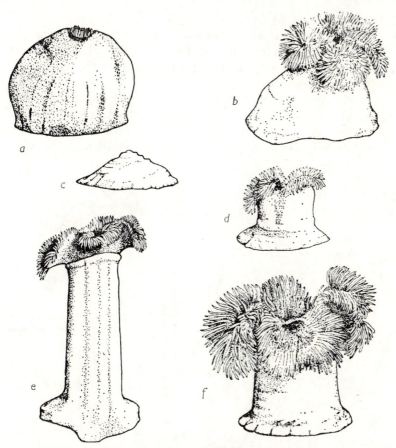

Figure 68. Outlines traced from photographs to the same scale of the same individual of *Metridium* on different occasions. (From Clark, 1964, after Batham and Pantin, 1950a)

muscle antagonism. The term is also applied to the parenchyma of platyhelminths and other tissues which work in the same way, even though one would not normally think of them as liquids.

We used as an example a cylindrical animal with longitudinal and circular muscles. Not all hydrostatic skeletons work in quite this way, as we shall see when we consider the foot of bivalve molluscs. Not all are permanently closed cavities, as we shall see when we consider sea anemones.

Sea anemones

The sea anemone *Metridium* can make extraordinary changes of size and shape (Batham and Pantin, 1950a, b). Fig. 68 was traced from photographs to the same scale, of the same individual at different times.

A sea anemone is essentially a cylindrical bag of seawater (Fig. 69). Flaps known as mesenteries extend radially inward from its inner wall. Both the body wall and the mesenteries are made of jelly-like mesogloea (see page 91) sandwiched between

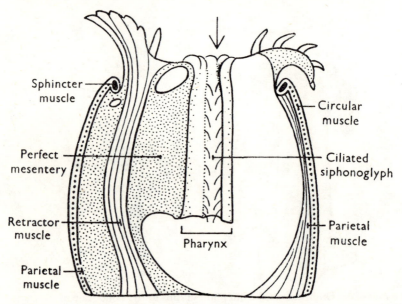

Figure 69. A diagrammatic section through a sea anemone (*Metridium*). (From Batham and Pantin, 1950a)

two layers of cells. There are circular muscles running round the body and longitudinal ones running up the mesenteries. The latter include parietal muscles close to the body wall and strong retractor muscles nearer the centre of the animal. The mouth is the only entrance to the body cavity (apart from some tiny pores), and is a slit in the top end of the animal, surrounded by the tentacles. At one or both ends of the slit there is a siphonoglyph, which is a tract of cilia which beat inwards. The siphonoglyphs pump water into the animal and can inflate it, if the rest of the mouth is kept closed.

We have already seen that the mesogloea is elastic so that it always returns to a particular size and shape when there are no forces acting on it (page 93). Anemones gradually acquire a particular size and shape when they are narcotized, and these are presumably the sizes and shapes at which the mesogloea is unstrained. The anemone in Fig. 68d is probably just a little smaller than it would be if it were narcotized. Figs. 68a, e and f show the same anemone greatly swollen, with the mesogloea stretched. Fig. 68c shows it contracted with the mesogloea compressed and so thickened, and also wrinkled. If the specimen were narcotized in any of these positions, the elasticity of the mesogloea would gradually restore it to the unstrained size and shape.

Increases of size are slow because the siphonoglyph cannot produce large pressures and the mesogloea is very viscous; for instance, it takes the animal an hour or more to lengthen to the shape shown in Fig. 68e. Contraction to the condition shown in Fig. 68c can be much faster, because the retractor muscles are quite strong. An anemone contracts in a few seconds, when it is poked.

The water-filled body cavity acts as a hydrostatic skeleton. If the mouth is kept closed the parietal and circular muscles are antagonistic. The anemone can change from the shape shown in Fig. 68e to the shape shown in Fig. 68f and back again, by contracting first the parietal muscles and then the circular ones. It can bend to one side and then the other, by contracting the parietal muscles of those sides in turn.

Fig. 70 shows apparatus used by Batham and Pantin (1950a) to measure the pressure inside living anemones. A water manometer

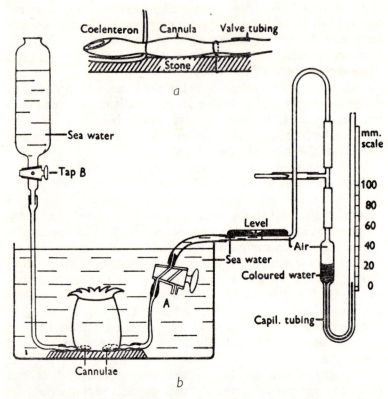

Figure 70. Apparatus for measuring the pressure in a sea anemone. (From
Batham and Pantin, 1950a)

is connected to a glass cannula which is slipped though a
slit in the body wall, into the body cavity. It is read with the tap A
in the position shown, and with the tap turned to connect it to
the open water. The difference between the two readings is the
difference in pressure between the contents of the body cavity
and the surrounding water. The funnel connected through tap B
to the second cannula was used in experiments in which the
anemone was inflated artificially by running in seawater, but
these need not concern us.

When the pressure in an anemone was observed over a period
of half an hour or so, marked fluctuations were found.

Anemones are for ever making small movements, and the pressure was found to rise sharply whenever the circular muscles contracted. The average pressure over a fairly long period varied between about 0·1 and 0·6 cm water, in different anemones. Rough calculations using equations (18) and (19) indicate that this is about the range of pressures that would be needed to counteract the elastic restoring force of the mesogloea, in moderately inflated anemones (Alexander, 1962).

Burrowing bivalves

A large proportion of bivalve molluscs live below the surface of sand or mud, with only their siphons protruding. They use cilia on their gills to draw water in through one siphon and drive it out through the other, and they filter this water to obtain particles of food. Some burrow deeply to escape from danger.

Fig. 71 is a side view of the razor shell, *Ensis*. The two long valves which form the shell are hinged together. The siphons project from between them at the top, and the foot projects at

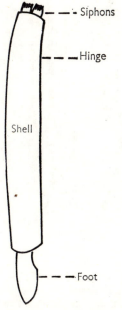

Figure 71. A sketch of the razor shell, *Ensis*

the bottom. (They can be withdrawn into the shell.) *Ensis* is an odd shape, but it burrows in the same way as other bivalves (Trueman, 1967). Its burrowing movements can be observed, if it can be persuaded to burrow just inside one wall of a transparent plastic box full of wet sand.

Fig. 72 shows how *Ensis* burrows. Specimens 13 cm long take about 2·5 s for each cycle of movements. At stage A in the cycle the shell is wide open, the siphons are open and the foot probes the sand. Since the open shell is jammed tightly in the sand, the foot penetrates the sand and scarcely pushes the shell

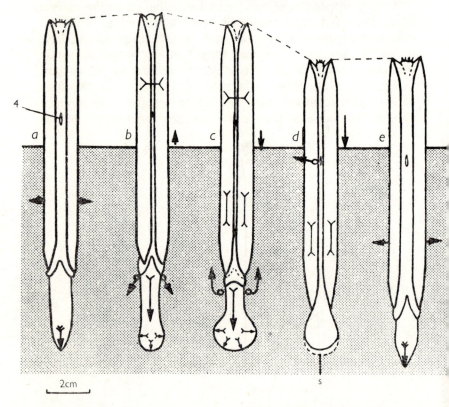

Figure 72. Diagrammatic ventral views of *Ensis* at five successive stages in the cycle of burrowing movements. >⟶ movements of fluid in the haemocoele; o⟶ ejection of water from the mantle cavity; >⟶< contraction of adductor or retractor muscles. (From Trueman, 1967)

up at all. In stage *b* the siphons are closed, the shell closes and squirts water out of its lower end, and the end of the foot swells. In stage *c* the foot shortens. Its swollen end is firmly anchored in the sand but the shell, since it has closed, is no longer tightly jammed. Therefore, the shell is pulled down as the foot shortens. At stage *d* the downward movement is finished. The siphons open and water is drawn in through them as the shell opens. The foot becomes less swollen. At stage *e* the animal is ready to start the cycle again.

If we are to understand how these movements are made we must know something of bivalve anatomy. Fig. 73 is a diagrammatic section through a typical bivalve, showing only the structures we need to know about.

The part of the shell near the hinge holds the visceral mass, with the adductor muscles which close the shell. The elastic hinge ligament opens the shell, when the adductor muscles relax (page 82). The foot protrudes from the visceral mass and the blood-filled body cavity (the haemocoele) extends into it. It has two main sets of muscles. There are retractor muscles from the shell to the end of the foot, which shorten it. There are transverse pedal muscles running across the haemocoele from one side of the foot to the other, which make the foot more slender. These two sets of muscles are antagonistic so long as the shell does not open or close, but stays in one position. When the retractor muscles contract the foot gets fatter as well as shorter, and when the transverse muscles contract it gets longer as well as thinner, since its volume does not change. However, when the shell closes it squeezes blood out of the visceral mass and inflates the foot, and when it opens it draws blood out of the foot into the visceral mass.

The shell encloses the mantle cavity as well as the visceral mass. This is the cavity in which the gills lie, through which the feeding current is drawn. It is of course filled with seawater. When the shell closes the volume of the mantle cavity is reduced as well as the volume of the visceral mass. Water must escape by the siphons or by some other route, such as through the gap between the foot and the edge of the shell (see the arrows A in Fig. 73). In *Ensis* the wall of the mantle cavity extends right across the gap between the two valves, and the only openings of

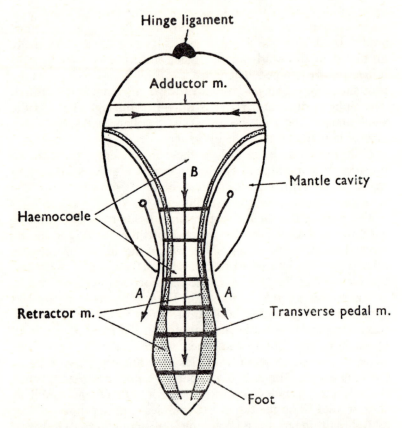

Figure 73. Diagrammatic transverse section of a typical bivalve. The arrows show movements of blood and water caused by contraction of the adductor muscles. (From Trueman, Brand and Davis, 1966)

the mantle cavity are the siphons, the gap round the foot and a small fourth aperture which is labelled 4 in Fig. 72. The siphons and the fourth aperture are closed at the stage in the digging cycle when the adductors contract, so all the water that is ejected comes out round the base of the foot, where it disturbs the sand and makes the shell's passage easier. Trueman (1967) estimated that when an *Ensis* whose volume was 30 ml adducted its valves, about 2 ml blood was driven into the foot and about 4 ml water was ejected from the mantle cavity.

Trueman (1967) measured the pressure in the haemocoele of burrowing *Ensis*. He stuck a hypodermic needle through the fourth aperture (Fig. 72) into the foot, and connected it by a short length of pressure tubing to a pressure transducer. The specimens burrowed normally until they were half buried and the tubing hit the surface of the sand. There were brief pulses of pressure of up to 100 cm water when the valves were adducted in each cycle of burrowing movements and much smaller pulses when the foot made its probing movements. The high pressures produced by the adductor muscles are needed to anchor the foot by making it swell. The adductors cannot be used to produce the probing movements because the shell has to be kept jammed in the burrow if the probing is to be effective. Probing is probably done entirely by the transverse pedal muscles.

Fig. 74 summarizes the observations on *Ensis*, and similar observations on other bivalves (Trueman, 1966a, b; Trueman, Brand and Davis, 1966). It shows the sequence of events in burrowing for a typical bivalve. It repeats a lot of the information shown in a different way in Fig. 72 but it also shows how the pressure rises at the same time as the valves are adducted, before the shell is pulled down into the sand.

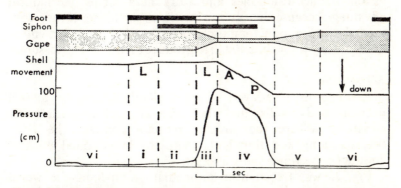

Figure 74. A diagram showing some of the events in a cycle of typical bivalve burrowing movements. A black bar opposite *foot* indicates that the foot is making probing movements and a hollow bar indicates that it is dilated. A bar opposite *siphon* indicates that the siphons are closed. The pressure in the haemocoele is given in cm water. (From Trueman, 1966a. Copyright © 1966 by the American Association for the Advancement of Science)

100 cm water is quite a high pressure. It is developed in the mantle cavity as well as in the haemocoele. What tension is needed in the adductor muscles to produce it? An *Ensis* 13 cm long has valves whose area in side view is about 25 cm² each (this is a projected area, see page 165). A pressure of 100 cm water (10^5 dyn/cm²) will exert a force of $2 \cdot 5 \times 10^6$ dyn on each valve, which will act half-way across the valve, 1 cm from the hinge. It will have a moment of $2 \cdot 5 \times 10^6$ dyn cm about the hinge, tending to open the shell. The elasticity of the hinge ligament also tends to open the shell, and this brings the total movement which the muscles would have to balance to about $2 \cdot 8 \times 10^6$ dyn cm. Trueman (1967) estimated that if the adductor muscles alone closed the shell and produced the pressure, they would have to develop a force of 5×10^6 dyn/cm² of cross-sectional area. This is within the bounds of possibility though it is about equal to the biggest force that has been recorded in experiments with bivalve adductor muscles and twice as high as the force frog muscle can produce (page 107). However, there are muscles which may help the adductors, which run between the edges of the valves in the wall of the mantle cavity. They have about as large a cross-sectional area as the adductor muscles and work at a higher mechanical advantage because they are further from the hinge. If they contracted with the adductor muscles, a force of $1 \cdot 5 \times 10^6$ dyn/cm² in both muscles would be enough to produce the observed pressure.

Earthworms

Earthworms have layers of longitudinal and circular muscle in their body walls, and their coeloms act as hydrostatic skeletons. Where the circular muscles contract the body becomes long and thin and where the longitudinal muscles contract it becomes short and fat.

Fig. 75 was traced from a cinematograph film of a worm crawling. In each segment of the body, the circular and longitudinal muscles contract in turn. In each segment the contractions happen very slightly later than in the segment in front of it. The effect is a series of waves of contraction of the circular muscles and of the longitudinal muscles, travelling alternately in a posterior direction along the body.

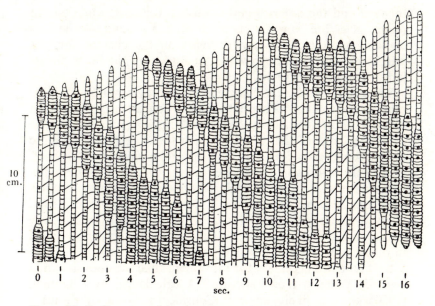

Figure 75. Outlines traced from photographs taken at 0·5 s intervals, of an earthworm crawling. (From Gray and Lissmann, 1938a)

Each segment has short, bristle-like chaetae which protrude from the body, pointing posteriorly. It is because of them that a worm feels rough if you run a finger forwards along its ventral surface, though it feels smooth if you run your finger backwards. Because of them, a worm can slide forwards over the ground more easily than backwards, and a worm could crawl simply by alternately extending and contracting its body. However, the chaetae are not fixed, but each segment protrudes its chaetae at the same time as it contracts its longitudinal muscles and becomes fat (Clark, 1964). While they are fat, the segments are stationary. Since the waves travel backwards along the body longitudinal muscles of the segments immediately posterior to a fat region are contracting, pulling the posterior parts of the body forwards. Similarly, the circular muscles of the segments at the anterior end of a fat region contract, lengthening the segments and pushing the anterior parts of the worm forward. The posterior end of the body is pulled forward and the anterior end is pushed forward, while the fat region gives the worm a purchase on the ground. While it is crawling on the surface, it depends on the chaetae for this purchase. When it is burrowing, the fatness of the fat segments jams them in the burrow.

The coelom of an earthworm is divided by septa into a large number of segments. The septa are not complete for each has a foramen round the nerve cord, but there is a sphincter muscle round the edge of the foramen which looks as if it could close tightly round the nerve cord and make the septum waterproof. Newell (1950) obtained evidence that the foramina are closed in active worms. He injected material opaque to X-rays into a few segments of some worms, took X-ray photographs of them, allowed them to wriggle for a while and then took X-ray photographs again. He found that the opaque material stayed in the segments where it was injected.

Septa are not necessary for movements like the ones shown in Fig. 75. What is their function? A calculation by Chapman (1950) gives a likely answer. Consider a worm of radius r whose circular and longitudinal muscles have areas A_C, A_L in a transverse section of the body.* Let the maximum force which can be developed per unit cross-sectional area of muscle be S. The longitudinal muscles can develop a force $A_L S$ and since the area of a cross-section of the worm is πr^2 they could produce a pressure $A_L S/\pi r^2$. The thickness of the layer of circular muscles is $A_C/2\pi r$ so the maximum tension (force per unit length of body) which they can produce is $A_C S/2\pi r$. We can work out the pressure they could produce from equation (18) (page 166): it is $A_C S/2\pi r^2$.

Hence, if maximal contractions of the circular and longitudinal muscles were to produce equal pressures, A_L should be about half A_C. In fact it is about five times A_C. The longitudinal muscles must be able to produce pressures many times larger than the circular muscles can.

If earthworms had no septa this would be a ridiculous arrangement. The pressure would always be the same all along the length of the worm and the worm could not make full use

---

* Chapman gave a different meaning to the symbol A_C.

of its longitudinal muscles. If the longitudinal muscles contracted maximally in a fat region of the body to jam it tightly in the burrow, the circular muscles in the thin parts could not contract and would indeed give way.

The argument so far is too simple. We have taken no account of the thinness of the thin regions: since they are thinner than the thick ones, less tension is needed to resist a given pressure (equation 18). On the other hand, we have taken no account of the thinning of the circular muscle layer by extension of the body. We have also taken no account of the way the force a muscle can produce changes as it shortens (Fig. 45, page 109). Still, A_L is so much greater than A_c that we can hardly escape the conclusion that a worm without septa could not use its longitudinal muscles fully.

The septa make it possible for the pressure to be different in different parts of the body. They probably cannot sustain much difference of pressure between one segment and the next, but quite substantial differences of pressure may be possible between segments some distance apart if each septum sustains a small difference. A maximal contraction of the longitudinal muscles in the middle of a fat region of the body need not be prevented by the weakness of the circular muscles in the thin regions.

One might think it desirable for the circular muscles to produce the higher pressures, for they are the muscles that thrust the worm's head forward through the soil. Chapman (1950) suggests however that worms may choose paths through existing crevices in the soil, and that the biggest forces may be needed for enlarging the crevices. These forces would be produced by the longitudinal muscles.

Newell (1950) measured the pressure in the coeloms of worms, using a device called a spoon gauge. He could detect no difference between atmospheric pressure and the pressure in anaesthetized worms but found excess pressures averaging about 15 cm water in the anterior segments of active worms and rather less in the posterior segments. The pressure fluctuated greatly between readings in each worm, and the highest pressure recorded was 29 cm water. This is presumably not greater than the maximum pressure the circular muscles can produce or withstand, for the worms were not burrowing. It is only in a burrow, where swelling is resisted by the surrounding soil, that the longitudinal muscles produce their maximum pressure.

Manton (1965) used a very simple method to measure the maximum pressure the longitudinal muscles can produce. She made a worm crawl under a microscope slide, set on supports so as to form a bridge, but with the supports so low that the worm had to raise the bridge to get under it. Weights were set on the bridge, until the worm could only just lift it. The weights were put at the sides of the bridge so that the worm could be photographed from above, through the glass of the loaded bridge. The area of the body in contact with the glass was measured in the photograph. If this area was A cm² when a weight of W g was being lifted, the pressure in the segments under the bridge must have been Wg/A dyn/cm². Manton found that the earthworm *Allolobophora* could produce pressures up to about 10^5 dyn/cm² or 100 cm water. This pressure was apparently produced by the longitudinal muscles, for the segments in contact with the bridge had their longitudinal muscles contracted. The measurements were rather rough, because the area of contact could not be distinguished very clearly in the photographs. It could have been measured more accurately if a prism had been used for the bridge, in the way Hammond (1966) used a prism in measuring pressure in *Acanthocephalus*.

Gray and Lissmann (1938b) used the apparatus shown in Fig. 76 to measure the longitudinal forces produced by a crawling earthworm. P,P is a fixed platform with a gap in it. B is a bridge mounted so that it can move a little forward and back, with weights W attached to it which tend to swing back to its original position if it is displaced. The more the bridge is displaced, the greater is the restoring force, so the position of the bridge at any moment indicates the force that is acting on it. Movements of the bridge are magnified by the lever p which writes on the smoked kymograph drum D. The weights are immersed in oil baths, of which only one is shown. The viscous oil damps the movements of the bridge and prevents it from going on swinging like a pendulum after every disturbance (see page 282).

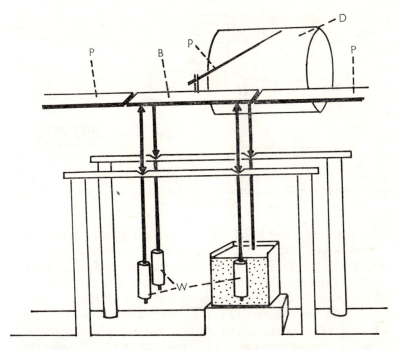

Figure 76. Apparatus for measuring forces exerted by crawling earthworms. The labels are explained in the text. (Redrawn from Gray and Lissmann, 1938b)

A worm was made to crawl along the platform and across the bridge. The forces acting on the bridge were recorded automatically on the smoked drum, and a cinematograph film was taken at the same time to show how they were produced. When a worm first reached the bridge (Fig. 77*a*) it had an anchored, thick region on the fixed platform, and it pushed its anterior segments forward against friction over the bridge. The bridge was pushed in the direction of movement. Later (Fig. 77*b*) there was a fixed region on the bridge, which pushed segments forward over the platform in front and dragged segments forward over the platform behind. The bridge was moved backwards. As the worm left the bridge (Fig. 77*c*), the bridge was pulled in the direction of movement again.

The force acting on the bridge when the worm was in the

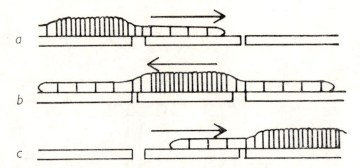

Figure 77. Diagrams of an earthworm crawling towards the right on the apparatus of Fig. 76. The arrows show the direction in which the movable platform moves

position shown in Fig. 77*a*, is the forward thrust developed by contraction of the circular muscles. The greatest value recorded was 8 g wgt (8,000 dyn). From this we can calculate the pressure in the segments where the circular muscles were contracting (Chapman, 1950). The cross-sectional area of these segments must have been about $0 \cdot 3$ cm² so the pressure must have been about 8,000/$0 \cdot 3$ dyn/cm², or about 27 cm water. This agrees well with the maximum pressure of 29 cm water which Newell (1950) measured with his spoon gauge. It is of course possible that the circular muscles might be able to produce more pressure than this, if the worm was pushing against a barrier.

The measurements that have been made show that the circular muscles can produce pressures of at least 29 cm water and that the longitudinal ones can produce pressures of 100 cm water. The calculation based on the areas of the muscles in transverse section led us to expect the longitudinal muscles to produce the greater pressure, though it suggested an even greater disparity in pressure.

Pressure and work

Think of a piston in a pump. Suppose its cross-sectional area is A and that it forces a volume V of fluid out of the pump at a pressure P. The force on the piston is PA. The distance it moves is V/A. Therefore the work it does is $PA \times V/A$ or PV. This is true of other sorts of pump as well as piston pumps. The work done by a pump is the volume of fluid pumped multiplied by the difference in pressure between the outgoing fluid and the incoming fluid.

Spiders' legs

The legs of spiders have hinge joints which have flexor muscles but no extensor muscles. The joints are extended by driving fluid into them under pressure (Parry and Brown, 1959a).

Fig. 78 shows a method for measuring the pressure in a spider's leg. One leg of a living spider is sealed into a glass tube connected to a manometer and, by a stopcock, to a compressed air supply. The flexible articular membrane of a joint is watched through a microscope and the pressure in the tube is increased gradually by admitting compressed air until the articular membrane collapses and becomes concave instead of convex. The pressure in the tube then just exceeds the pressure in the leg. When the spider is inactive this pressure is about 5 cm mercury above atmospheric pressure.

The pressure rises briefly to much higher values when the

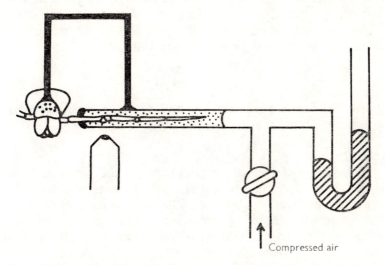

Figure 78. Apparatus for measuring the pressure in a spider's leg. (Redrawn from Parry and Brown, 1959)

spider is stimulated by touch or by a current of air. Parry and Brown used a pressure transducer to measure this pressure in some experiments, but were also able to measure it with the simple apparatus shown in Fig. 78. They raised the pressure in the tube to a selected value above the resting pressure, so that the articular membrane collapsed, and then stimulated the spider. If the membrane re-inflated, the pressure in the leg must have risen above the pressure in the tube. They found that struggling spiders could develop pressures up to 45 cm mercury (0·6 atm) but could only maintain them for a fraction of a second.

When two legs were enclosed in tubes it could be shown that the pressures in them changed simultaneously and by the same amount. This suggests that the pressures are produced in the spider's body. They are probably produced by muscular contraction of the prosoma (the anterior half of the body).

Pressure extends the joint because its volume is greater when it is extended than when it is flexed. The work done by the muscles of the body, driving fluid into the joint, is the product of the pressure P and the change in volume of the joint, ΔV. The work done by a joint as it straightens against a resisting force is the product of the moment of this force about the joint, M, and the change in the angle of the joint, $\Delta\theta$ (see page 28). If the cuticle of the leg is not stretched appreciably by the pressure no appreciable work will be used in stretching it and

$$P.\Delta V = M.\Delta\theta$$
$$M/P \ = \Delta V/\Delta\theta$$

Parry and Brown measured M/P and $\Delta V/\Delta\theta$ in separate experiments and found that they were in fact about equal.

Density

The density of a substance is its mass per unit volume. When a density is given, the units (e.g. g/cm³) must be stated. The specific gravity of a substance is the ratio of its density to the density of water at 4°C. Since it is a ratio it has no units. The density of water is 1·000 g/cm³ so the specific gravity of a substance is numerically equal to its density in g/cm³.

The density of a liquid can be measured by means of a density bottle or a hydrometer, provided that a reasonable quantity of the liquid is available. The methods are described in elementary books on physics (e.g. Nelkon and Parker, 1965) which also describe how a density bottle can be used to determine the density of solid objects. It can be used, for instance, to determine the density of fish scales which are available in fairly large numbers and are small enough to pass through the neck of a density bottle (Alexander, 1959b).

The densities of small samples of tissues, of small animals or of drops of fluids like blood can be determined by making a mixture of chloroform and benzene of the same density as the sample, and measuring the density of the mixture with a hydrometer (see, for instance, Alexander, 1959b). The density of benzene is $0 \cdot 90$ and the density of chloroform is $1 \cdot 49$, so mixtures can be made of any density between these limits. Outside the limits, other liquids would have to be used. The specimen is put in a mixture of the liquids and one liquid or the other is added until, as far as possible, the specimen neither floats nor sinks. It is necessary to stir very thoroughly each time liquid is added and to avoid getting air bubbles stuck to the specimen.

The density of large animals has sometimes been determined by weighing them and measuring their volume. The volume is measured by putting the animal into a jar filled to the brim with water and measuring the volume of water that overflows. This method is inaccurate. The volume could scarcely be measured with an accuracy better than 1% so one could not, for instance, decide with any confidence whether the density of a fish was $1 \cdot 07$ or $1 \cdot 08$ g/cm³.

A much better method of measuring the density of a reasonably large animal uses Archimede's Principle: that when an object is immersed in a fluid an upward force acts on it, equal to the weight of fluid it displaces. Consider an object of weight W and density ρ immersed in a fluid of density ρ'. The volume of the object is W/ρ so the weight of fluid it displaces is $W\rho'/\rho$. The net downward force on it, W', is given by the equation

$$W' = W - W\rho'/\rho$$

Whence
$$\rho'/\rho = (W - W')/W$$
$$\rho = \rho'[1 + W'/(W - W')] \tag{21}$$

W can be determined by weighing the object in air in the

usual way. W' can be determined by weighing it as it hangs immersed in water or some other fluid of known density ρ'. Hence ρ can be calculated. An error of 1% in either W or W' will not cause a 1% error in ρ, but only in the difference between ρ and ρ'. In the case of a fish of density $1\cdot070$ g/cm³ weighed in air and water, it would only alter the result to $1\cdot069$ or $1\cdot071$. This method has an enormous advantage over the method of measuring weight and volume, for objects whose densities are close to the density of water.

Even this method may be unsatisfactory for determining the densities of fairly small aquatic animals, such as prawns. Such animals are impossible to weigh at all accurately in air, because of the difficulty of drying the surface of the animal without desiccating it. Lowndes (1942) has devised a rather elaborate method of determining density which avoids this difficulty.

An animal consists of a collection of tissues of different densities. For instance, in the case of fish, muscle has a density of about $1\cdot05$ g/cm³, bone (when dry) about $2\cdot0$, cartilage about $1\cdot1$ and fat about $0\cdot9$ (Lowndes, 1955; Alexander, 1959b). The density of the whole animal depends on the proportions and densities of its components. Consider an animal composed of a volume x of material of density ρ_x and a volume y of material of density ρ_y. The weights of the two components will be $x\rho_x$ and $y\rho_y$, respectively, and the density ρ of the whole animal will be given by

$$\rho = (x\rho_x + y\rho_y)/(x + y) \tag{22}$$

Buoyancy of cuttlefish

The cuttlefish *Sepia* hovers near the bottom of the sea, stalking small fish, shrimps and crabs and shooting out its long tentacles to catch them. It needs little energy to hover in mid-water, because its density is very close to the density of sea-water. Denton and Gilpin-Brown (1961a) studied the buoyancy of cuttlefish.

They determined the density of anaesthetized cuttlefish by weighing them in air and in seawater. Some were less dense than seawater, and bits of lead had to be tied to them to make them sink. The weight of the animal in seawater was of course negative in these cases and was the difference between the weight in seawater of the animal and lead, and that of the lead alone. The weight in seawater of different cuttlefish varied between about $+0.4\%$ and -0.9% of the weight in air. The density of the seawater was 1.026 g/cm³ and, by equation (21), the densities of the cuttlefish must have ranged from 1.017 to 1.030 g/cm³.

Cuttlefish owe their low density to the cuttlebone whose shape and position are shown in Fig. 79. It consists of a stack of about a hundred thin chambers partly filled with gas, sandwiched between partitions made of chitin impregnated with calcium carbonate. The broken lines across the cuttlebone in the figure represent only some of the partitions. The partitions are held apart by pillars made of the same material as themselves. Each chamber is completely enclosed by the calcified chitin.

Cuttlebones were excised from cuttlefish the densities of which had been determined, and the densities of the cuttlebones were then determined by weighing in air and seawater. They had to be loaded with lead to make them sink. Their densities averaged 0.62 g/cm³.

The cuttlebone occupies about 9.3% of the volume of the

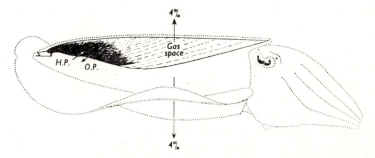

Figure 79. Outline of a cuttlefish (*Sepia*) showing the position of the cuttlebone and (in black) the distribution of water in the cuttlebone. The arrows indicate that the buoyancy of the cuttlebone balances the weight in sea water of the rest of the animal and that the hydrostatic pressure difference between the sea and the interior of the cuttlebone is balanced by the osmotic pressure difference between the cuttlebone liquid and the blood. (From Denton and Gilpin-Brown, 1961b)

animal. Let us work out the density the animal would have without it. A typical cuttlefish would have a density about the same as seawater ($1 \cdot 026$ g/cm³) and would consist of about $9 \cdot 3$ parts by volume of cuttlebone of density $0 \cdot 62$ g/cm³ and $90 \cdot 7$ parts by volume of other material whose density we want to know. By equation (22)

$$1 \cdot 026 = (9 \cdot 3 \times 0 \cdot 62 + 90 \cdot 7\, \rho'')/100$$
$$\rho'' = 1 \cdot 067 \text{ g/cm}^3$$

If the cuttlefish had no cuttlebone it would, therefore, be about 4% denser than seawater, like the squid *Loligo* which has a vestigial cuttlebone without gas spaces.

Though the chambers of the cuttlebone are filled largely with gas they also contain some fluid. Denton and Gilpin-Brown found out how much fluid there was, in cuttlebones dissected from anaesthetized cuttlefish. First they weighed the cuttlebone in air and water and determined its density. Then they stuck a needle through it several times so as to puncture all the chambers, and put it in a dish of seawater in a vacuum chamber. As the pressure was reduced, gas bubbled out of the chambers. When no more could be extracted they restored atmospheric pressure, and the cuttlebone filled with seawater. They weighed it full of seawater, and then dried it in an oven till all the water had evaporated and weighed it again. They give an example of their results:

Initial weight of cuttlebone in air	$30 \cdot 5$ g
Initial density	$0 \cdot 68$ g/cm³
Weight full of seawater	$55 \cdot 1$ g
Weight after drying	$17 \cdot 5$ g

The dried cuttlebone was not empty but contained the salt left behind by the evaporated seawater. $37 \cdot 6$ g water evaporated so about $1 \cdot 3$ g salt must have been left. The weight of the dry cuttlebone without salt would be $16 \cdot 2$ g and the weight of seawater which filled the chambers $55 \cdot 1 - 16 \cdot 2 = 38 \cdot 9$ g. The weight of fluid the cuttlebone contained initially must have been $30 \cdot 5 - 16 \cdot 2 = 14 \cdot 3$ g and we will assume it had the same density as seawater. Thus we find

Volume of complete cuttlebone $= 30\cdot5/0\cdot68$ $= 45$ ml
Volume of chambers $= 38\cdot9/1\cdot026 = 38$ ml
Volume of fluid initially in chambers $= 14\cdot3/1\cdot026 = 14$ ml
Volume of gas initially in chambers $= 38 - 14$ $= 24$ ml
Volume of solid material $= 45 - 38$ $= 7$ ml

Thus, the chambers of this particular rather dense cuttlebone were a little more than one-third full of fluid and a little less than two-thirds full of gas. The density of the solid material of the cuttlebone was $16\cdot2/7 = 2\cdot3$ g/cm³, which lies in the range of values one would expect for mixtures of chitin and calcium carbonate.

We will see later how cuttlefish control the densities of their cuttlebones (page 192).

Centre of buoyancy

In the same way as the weight of a body acts at the centre of gravity, the upthrust due to a weight of fluid displaced by it acts at a point known as the centre of buoyancy. If the body is completely submerged in a fluid the centre of buoyancy is where the centre of gravity would be, if the body was made of material of uniform density. The centre of buoyancy of a sphere, for instance, is always at the centre of the sphere. This is true even if the sphere is made half of lead and half of wood, although the centre of gravity of such a sphere would be well away from the centre, in the lead half. When the centre of gravity and the centre of buoyancy of an object are not in the same place, the weight acting at the centre of gravity and the upthrust acting at the centre of buoyancy tend to rotate the object so as to bring the centre of buoyancy directly above the centre of gravity. This is the basis of a method that has been used to locate the centre of buoyancy of *Nautilus*.

Buoyancy of Nautilus

Nautilus is a cephalopod whose shell looks very different from the cuttlebone of *Sepia* (Fig. 79) but is homologous with it and gives buoyancy in the same way. The animal and its coiled shell are shown in section in Fig. 80*a*. The chambers of the shell are filled with gas but the walls of the shell are quite thick and the

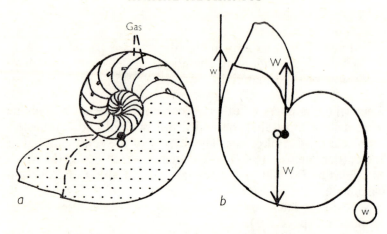

Figure 80. (*a*) A saggittal section of *Nautilus* showing the body (stippled), the gas chambers and the centres of gravity (○) and buoyancy (●). (*b*) The method of locating the centre of buoyancy

densities of shells taken from anaesthetized animals are about 0·95—much higher than the densities of cuttlebones (Denton and Gilpin-Brown, 1966). Nevertheless, the shell is big enough to make the animal very nearly the same density as seawater.

Nautilus floats in the position shown in Fig. 80*a*, with the gas-filled chambers above the body. Denton and Gilpin-Brown (1966) located its centres of gravity and buoyancy. In the natural position the centre of buoyancy must be directly above the centre of gravity. When the animal is rotated through 90° the centres must lie in the same horizontal plane. An animal in a bucket of seawater was arranged in this position, by means of threads attached to diametrically opposite points on its shell by rubber suckers. One thread was fixed to a support above the water and weights were hung from the other (Fig. 80*b*). The weight *w* needed to hold the animal at 90° to its natural position was found. If the weight of the animal is *W* and its density is the same as that of the seawater, the upthrust due to the weight of water displaced will also be *W*. If the centres of buoyancy and gravity are a distance *x* apart, these two forces *W* exert a couple *Wx* on the animal. The thread attached to the support must exert a force *w* on the animal, to balance the force *w* in the other thread, and if the points of attachment of the threads to the shell are a distance d apart the two forces w will exert a couple wd on the animal. Since the animal is in equilibrium

$$Wx = wd$$
$$x = wd/W$$

By this method it was found that the centres of buoyancy and gravity were 0·6 cm apart in a *Nautilus* which measured 13 cm across. The centre of gravity was found by the method described on page 20 to lie in the position shown in Fig. 80a. The centre of buoyancy must be in the position shown, 0·6 cm from the gravity and directly above it in the position in which the animal floats.

Osmotic pressure

Consider two solutions of the same solute dissolved in the same solvent in different concentrations. Let them be separated by a membrane which allows molecules of the solvent to pass through, but not molecules of the solute. Solvent will pass through the membrane from the more dilute solution to the more concentrated one, tending to equalize their concentrations. Its passage can be prevented, by applying a greater pressure to the concentrated solution than to the dilute one. The pressure needed is called the osmotic pressure. Nearly every biology student must at some time have seen a demonstration of osmotic pressure, involving a thistle funnel with a piece of parchment tied over its mouth. Some sugar solution is put in the funnel and it is inverted in a dish of water. Water passes through the parchment into the funnel, and continues to do so till the level of liquid in the funnel or in a tube attached to it is far above the level in the dish.

The osmotic pressure Π between a dilute solution and the pure solvent is given by the equation

$$\Pi = RTc \qquad (23)$$

where R is the universal gas constant (see page 193), T is the absolute temperature and c is the molar concentration of the solution (that is, the concentration in gram molecules per unit volume). If c is given in mol/cm³ and R in erg/° mol (it is $8 \cdot 3 \times 10^7$ erg/° mol), Π is obtained in dyn/cm².

Osmotic pressures are often very large, and so hard to measure directly. For instance, the osmotic pressure of sea-water is about 25 atm (Hale, 1965) and immensely strong apparatus would be needed to measure it. However, osmotic pressure is proportional to depression of the freezing point, which is easier to measure. The osmotic pressure at 0°C of an aqueous solution which freezes at $-\Delta T$°C is about 12 ΔT atm. Osmotic pressure can also be calculated from the boiling point or from the vapour pressure (Glasstone and Lewis, 1960).

Osmotic pressure is often discussed in biology, but usually in contexts which are generally classed as physiology rather than as animal mechanics.

Osmotic mechanism of the cuttlebone

Osmotic pressure seems to play a vital part in controlling the density of cuttlefish. Denton, Gilpin-Brown and Howarth (1961) studied cuttlefish trawled from a depth of 73 m. Since the density of water is 1 g/cm³, pressure under water increases by 10^5 dyn/cm² or $0 \cdot 1$ atm for every metre descent. Hence, the pressure 73 m below the surface would be $8 \cdot 3$ atm.

Fluid was extracted from the cuttlebone, and its osmotic pressure was determined either from the freezing point or from the vapour pressure. If the fluid was extracted as soon as possible after the animal had been brought to the surface its osmotic pressure was found to be about 18 atm. If it was extracted a few hours later the osmotic pressure was 24 atm or about the same as seawater. The osmotic pressure of the blood was about 24 atm all the time. Before the cuttlefish were caught the osmotic pressure of the fluid in their cuttlebones must have been at least 6 atm below the osmotic pressure of their blood. The difference would probably have been more than 6 atm, since the osmotic pressure of the cuttlebone fluid rises after capture. The difference would tend to draw fluid out of the cuttlebone. It must be balanced in life by a difference of hydro-static pressure tending to drive fluid into the cuttlebone (Fig. 79).

Denton and Gilpin-Brown (1961b) measured the pressure in the chambers of cuttlebones, through a hypodermic needle attached to a manometer. The pressures they found were below atmospheric pressure. The average was $0 \cdot 8$ atm but the most recently formed chambers at the anterior end of the cuttlebone had pressures as low as $0 \cdot 3$ atm. It seems probable that when a chamber is formed the osmotic pressure of the fluid in it is reduced and that this causes fluid to pass out into the blood, leaving a partial vacuum. Gases subsequently diffuse into the chamber until the gases in it are in equilibrium with the gases dissolved in the animal's body fluids.

The gases dissolved in the blood of a cuttlefish tend to reach equilibrium with the gases dissolved in the seawater, which tend to reach equilibrium with the atmosphere. The partial pressure of the gases dissolved in the body fluids (i.e. the pressure of the mixture of gases with which they would be in equilibrium) cannot exceed 1 atm and is generally less than 1 atm because oxygen gets used for respiration. This is true, whatever the depth in the sea. Even if the cuttlefish lives at 73 m where the pressure is $8 \cdot 3$ atm, gas will only diffuse into the cuttlebone until the pressure in it is 1 atm or a little less. The remaining pressure difference of $7 \cdot 3$ atm or a little more is balanced by the difference in osmotic pressure between the cuttlebone fluid and the blood. When the cuttlefish is brought to the surface the hydrostatic pressure difference becomes very small and the osmotic pressure difference is reduced accordingly.

The cuttlebone has to be strong, to withstand the pressure difference between its contents and its surroundings. Cuttlefish live at depths down to 150 m where the pressure is 16 atm so that the pressure difference is over 15 atm. Denton, Gilpin-Brown and Howarth (1961) enclosed a cuttlebone in a close-fitting polythene bag to prevent water from seeping in, and then put it in a pressure chamber and increased the pressure gradually. It did not collapse until the pressure reached 24 atm.

The ideal gas equation

One mole of an ideal gas at pressure P and absolute temperature T occupies a volume V such that

$$PV = RT \tag{24}$$

This is the ideal gas equation, and R is the universal gas constant. If the pressure is in dyn/cm² and the volume in cm³, R must be given its value in erg/° mol, which is $8 \cdot 3 \times 10^7$ (Glasstone and Lewis, 1960). Boyle's law follows from the ideal gas equation: if the volume of a given mass of gas is V_1 at a pressure P_1 and V_2 at a pressure P_2 at the same temperature

$$P_1 V_1 = P_2 V_2 \qquad (25)$$

The ideal gas equation describes the behaviour of real gases sufficiently accurately for most purposes, except at very high pressures. An example will show its limitations. The density of oxygen at atmospheric pressure and ordinary temperatures is $1 \cdot 4 \times 10^{-3}$ g/cm³. The equation predicts that at 700 atm it should occupy 1/700 of its volume at 1 atm, and that its density should be $700 \times 1 \cdot 4 \times 10^{-3} = 1 \cdot 0$ g/cm³. Its density at this pressure is in fact $0 \cdot 7$ g/cm³ (National Research Council, 1928). A fish with a gas-filled swimbladder has been caught at a depth of about 7,000 m where the pressure is about 700 atm, and the difference between real and ideal gases had to be considered in a discussion of the effect of the gas on the density of the fish (Alexander, 1966b). This was an exceptional case. For most zoological purposes gases can be regarded as ideal gases.

Swimbladders

Many teleost fish have densities very close to the density of the water they live in, and can hover in mid-water with only very gentle fin movements. This is because they have a bag of gas called the swimbladder in the body cavity. Without it, their densities would lie between about $1 \cdot 06$ and $1 \cdot 09$ g/cm³. They would sink whenever they stopped swimming and they would need more energy to swim at any given speed. It has been estimated that a fish which swims all the time is likely to use 8 % less energy in its metabolism if it has a swimbladder, than if it has not (Alexander, 1967b).

The pressure of the gas in a cuttlebone is less than the pressure of the surrounding water (page 193). The pressure of the gas in a swimbladder is the same as the pressure of the surrounding water or (as we shall see) slightly more. Cuttlefish reduce their density by withdrawing fluid from a container with rigid walls. Fish do it by secreting gas into a container with flexible walls (for an account of the mechanism of secretion see Alexander, 1966b). The outer wall of a teleost's body is flexible, as well as the wall of the swimbladder itself, so the gas in the swimbladder is compressed as the fish swims deeper and expands as it swims towards the surface. With a given quantity of gas in it there is only one depth at which the swimbladder has exactly the right volume to make the fish the same density as the water. Even at this depth the equilibrium of the fish is unstable, for a slight rise will allow the swimbladder to expand, reduce the density of the fish and tend to make it rise further, while a slight fall below the equilibrium depth will increase the density of the fish and tend to make it sink further. The swimbladder is much less satisfactory in this respect than the cuttlebone, for the volume of the gas spaces in a cuttlebone is scarcely affected by changes of pressure (Denton, Gilpin-Brown and Howarth, 1961).

Fig. 81 shows apparatus for studying the effect of pressure changes on the volume of fish with swimbladders (Alexander, 1959a). The fish is anaesthetized and put in the flask, which is filled completely with weak anaesthetic solution. The solution extends part way along the capillary tube where it forms a

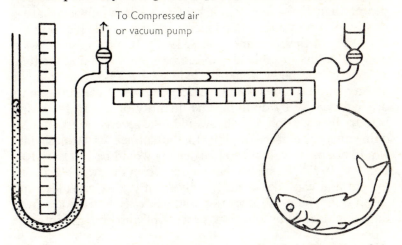

To Compressed air
or vacuum pump

Figure 81. Apparatus for investigating the mechanical properties of the swimbladders of fish. (Modified from Alexander, 1959a)

meniscus. The pressure is altered by means of compressed air or a vacuum pump and measured by the manometer. The distance the meniscus moves along the capillary is measured. The corresponding change of volume can be calculated, if the bore of the capillary is known. Various precautions have to be taken to avoid errors and to correct for them.

If the weight of the fish and its density at one known pressure are known, the density at any other pressure within the range of the experiment can be calculated from the volume changes. One way of getting the information about density is by finding the pressure at which the fish has exactly the same density as water. This can be done in the apparatus shown in Fig. 81, if the fish is not too tight a fit in the flask, by adjusting the pressure until the fish just floats (Alexander, 1959b).

From the results of the experiments it seems that the volumes of most teleost swimbladders vary with external pressure almost exactly as Boyle's law predicts (Alexander, 1959c). The walls of the swimbladder and of the body scarcely affect the changes of volume. A fish living in shallow water will commonly have almost exactly the same density as the water, but if it swam to a depth of 10 m where the pressure was 2 atm the swimbladder would be compressed to half its size and it would be a good deal denser than the water. Perch kept in shallow aquaria have a density of, on average, $1 \cdot 005$ g/cm³ and a swimbladder occupying $7 \cdot 5\%$ of their volume (Jones, 1951). If such a perch swam down 10 m the swimbladder would be compressed and the total volume of the body would be reduced by $3 \cdot 75\%$ so that the density became $1 \cdot 005 \times 100/96 \cdot 25 = 1 \cdot 045$. It could regain its original density by doubling the quantity of gas in the swimbladder by secretion, but this would take several hours. If after doing this it returned rapidly to the surface its swimbladder would double in volume and its density would be $1 \cdot 005 \times 100/ 107 \cdot 5 = 0 \cdot 935$. Some fish have a duct from the swimbladder to the gut and can release gas rapidly from the swimbladder and spit it out. The perch does not, and can only remove gas from the swimbladder by the slow process of absorbing it into the blood.

A fish which is not quite the same density as the water can generally still hover in mid-water, by moving its fins. The greater the difference in density the more power is needed for the fin movements, and the power could be estimated in the same way as the power consumption of a hovering humming bird is estimated on page 250. If the difference becomes too great, the fish is unable to hover. Jones (1952) found that perch were unable to hover when their density fell below $0 \cdot 98$ g/cm³. A perch adapted to a depth of 10 m could not hover near the surface or even at a depth of 4 m until it had removed some of the gas from the swimbladder.

There is a very large group of teleosts which do not have the usual highly extensible swimbladder walls, but rather inextensible ones which are taut because the swimbladder is inflated under slight pressure. This is the order Cypriniformes, which includes the carps and minnows and a large proportion of the other freshwater fish of the world. The volumes of their swimbladders do not change as much with depth as Boyle's law would lead one to expect, if one considered only the volume of the swimbladder and ignored the elasticity of its wall (Alexander, 1959a, 1961). These fish have been investigated by means of the apparatus shown in Fig. 81. The technique depends on their having a duct from the swimbladder to the gut. The duct is normally closed by a sphincter muscle but if the pressure is reduced beyond a certain point the duct opens and allows gas to escape.

Fig. 82 shows the results of an experiment on a rudd (*Scardinius*). In the first part of the experiment the pressure in the flask was increased by stages to 24 cm mercury above atmospheric pressure and the position of the meniscus at each pressure was noted. It was then reduced by stages to 6 cm mercury below atmospheric pressure. No gas was released from the swimbladder at this pressure so it was possible to draw a graph of change of swimbladder volume (indicated by meniscus position) against pressure, for this whole range of pressures. This is the graph whose points are shown as crosses in Fig. 82. At the start of the second part of the experiment the pressure was reduced well below atmospheric pressure so that a good deal of the gas escaped from the swimbladder as bubbles which floated up to the top of the flask. When atmospheric pressure was restored the volume of the swimbladder gas was greater than

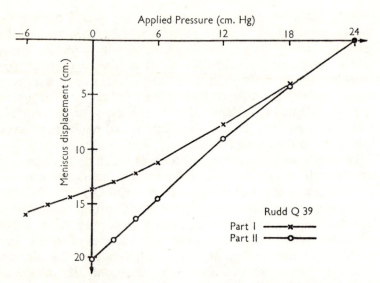

Figure 82. The results of an experiment with a rudd (*Scardinius*) in the apparatus of Fig. 81. (From Alexander, 1959a)

before, because it was no longer confined by a taut swimbladder wall. Enough gas had escaped from the swimbladder to make its wall slack. The pressure was increased again by stages to 24 cm mercury above atmospheric pressure, and the position of the meniscus was noted at each pressure. These positions are shown by circles in Fig. 82.

We now have two graphs, one showing the effect of external pressure on the volume of the gases confined in the swimbladder and the other showing the effect of pressure on the volume of the gases released from the swimbladder or at least no longer restrained by its wall. From the first graph we can calculate the effect of pressure changes on the density of the fish. From the second, using Boyle's law, we can calculate the quantity of gas in the swimbladder. From the two graphs we can read off the pressure in the swimbladder. In the first part of the experiment the meniscus was at 13 cm when the pressure in the flask was atmospheric pressure. In the second part of the experiment the pressure had to be increased to 6·9 cm mercury above atmospheric pressure to bring the meniscus to the same position.

Therefore the pressure of the gas confined in the swimbladder must initially have been 6·9 cm mercury above atmospheric pressure.

The value of the tightly inflated swimbladder of Cypriniformes is illustrated by Fig. 83. The roach (*Rutilus*) is a member of the order whose swimbladder is inflated at about the same pressure as in rudd. The fifteen-spined stickleback (*Spinachia*) belongs to a different order and has a swimbladder with slack, highly extensible walls. The figure shows how the volume of roach and *Spinachia* swimbladders change with pressure, and compares them with a free bubble of air. There is very little difference between the volume changes of the *Spinachia* swimbladder and the bubble, except when the pressure is reduced so much that the swimbladder wall becomes taut and resists further expansion of the gas. The volume of the roach swimbladder changes less than half as much, at pressures near atmospheric pressure. When a roach changes its depth its density changes less than half as much as if it had an ordinary swimbladder with a slack extensible wall.

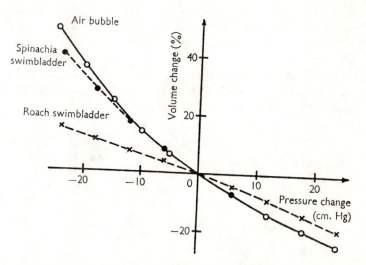

Figure 83. Changes of volume of an air bubble and of the swimbladders of *Spinachia* and roach (*Rutilus*), caused by changes of external pressure. (From Alexander, 1959c)

Surface tension

A molecule in a liquid, well away from the surface, is attracted equally in all directions by the surrounding molecules. A molecule at the surface is attracted only inwards. A liquid therefore tends to acquire a compact shape with the least possible surface area: it behaves in some respects as though its surface were an elastic skin. The surface is unlike an elastic skin, in that the tension in it is generally constant, however much it is extended or allowed to contract. The surface tension is the force acting at right angles to any line of unit length in the surface. It is equal to the surface energy, which is the work which must be done to increase the area of the surface by one unit. Surface tension is usually measured in dyn/cm and surface energy in erg/cm^2 but since an erg is a dyn cm these dimensions are identical.

The surface tension at a water surface exposed to the air is 73 dyn/cm at 20°C, and changes a little with temperature. Many organic liquids have surface tension of 30 dyn/cm or less, at surfaces exposed to the air. Surface tensions can also be measured at surfaces where two immiscible liquids meet.

When a liquid and air are in contact with each other and a solid surface, the surfaces meet at a particular angle which depends on both the liquid and the solid. It is known as the contact angle and is always measured through the liquid so that it is acute in Fig. 84a and obtuse in Fig. 84b. When the liquid is spreading over the solid surface the contact angle is greater than when it is retreating. The rougher the solid surface the more the contact angle differs from 90°, whether it be acute or obtuse.

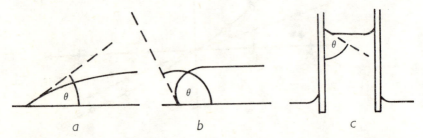

Figure 84. (a), (b) Side views of the edges of drops of liquid resting on flat solid surfaces. The contact angle θ is acute in (a) and obtuse in (b). (c) A liquid with an acute contact angle rising up a tube

Contact angles can be affected markedly by traces of impurity on the solid surface. The contact angle between water and glass is 0° if the glass is really clean, but may be as high as 8° if it is not quite clean (Nelkon and Parker, 1965). The contact angle between water and wax is about 105°.

A liquid rises up a tube which is dipped into it if the contact angle is acute (Fig. 84c) because the surface tension pulls it up.

Plastrons

Insects breathe gaseous air and those that live under water need special respiratory adaptations. The most remarkable of these is the plastron, found in the bug *Aphelocheirus* and a few beetles (Thorpe and Crisp, 1947, 1949; Crisp, 1964). It is a thin layer of gas covering much of the surface of the body and kept in place by large numbers of tiny hairs. Oxygen which is initially dissolved diffuses into the plastron and thence into the tracheal system which distributes it round the body. The plastron works as a sort of gill, and an insect with a plastron need never visit the surface for air.

If it is to last the plastron must be kept at a pressure below atmospheric pressure. This can be explained best by describing what happens in insects like *Corixa* which carry air underwater but do not have a plastron. Air contains about 20% by volume of oxygen and 80% of nitrogen, so the partial pressure of oxygen in the atmosphere is 0·2 atm and the partial pressure of nitrogen is 0·8 atm. Water which is standing exposed to the atmosphere tends to reach equilibrium with it, so that the partial pressures of oxygen and nitrogen dissolved in it are 0·2 atm and 0·8 atm, respectively. When *Corixa* carries air below the surface and removes some of the oxygen for respiration, the partial pressure of oxygen in this air falls below 0·2 atm and oxygen diffuses in from the water. Because the partial pressure of oxygen falls and because the total pressure of the submerged air is greater than 1 atm (since pressure increases with depth) the partial pressure of nitrogen rises above 0·8 atm. Nitrogen then diffuses out into the surrounding water. The oxygen content of the air store will never rise above a quarter of the nitrogen content, since diffusion from the water cannot raise the partial pressure of oxygen above 0·2 atm. Therefore, as nitrogen is lost the air store gets smaller and eventually *Corixa* has to visit the surface to collect more air.

This would not occur if the partial pressure of nitrogen were kept below 0.8 atm. The partial pressure of oxygen cannot rise above 0.2 atm, so the total pressure of the air store would have to be kept below 1 atm, at whatever depth the insect was swimming. *Aphelocheirus* lives in the River Volga at depths down to at least 7 m. The pressure at this depth is 1.7 atm and the air store in the plastron can only last if it is kept at a pressure 0.7 atm or more below the pressure of the surrounding water. We shall see how this is achieved.

The plastron of *Aphelocheirus* is a pile of tiny hairs standing more or less at right angles to the cuticle, but bent over at their tips (Fig. 85*a*). The air is trapped between the cuticle and the tips of the hairs. Its volume is kept almost constant irrespective of depth in the manner indicated in Fig. 85*b*, *c*. These are sections through the tips of a few adjacent hairs, cut in a plane at right angles to the plane of Fig. 85*a*. Fig. 85*b* shows how the air-water surface runs at small depths. It meets each hair at the contact angle θ. Its surface tension pulls on the hairs as indicated by the arrow. The resultant force on each hair acts downward, towards the cuticle. The hairs therefore exert an upward force on the surface which is balanced by a small pressure difference between the water and the air.

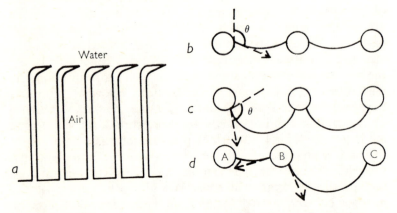

Figure 85. Diagrams of a plastron which are explained in the text

Fig. 85c shows what happens at a greater depth. The increased pressure has driven the water surface very slightly nearer the cuticle. The surface meets the hairs at the same contact angle as before but the forces that surface tension exerts on the hairs are now inclined much more steeply down towards the cuticle. The resultant force on each hair is greater and the force exerted by the hairs on the surface is greater and balances a greater pressure difference.

Thorpe and Crisp (1947) measured the size and spacing of the hairs in the plastron of *Aphelocheirus*, and estimated the maximum pressure difference that could be maintained between the air and the water. They were unable to measure the contact angle, but guessed that it would be about 110°. The advancing contact angle between smooth insect cuticles and water is usually 100–110° (Holdgate, 1955). They estimated that pressure differences up to 3·7 atm should be possible if the hairs were strong enough to stand them. This is far more than the greatest pressure difference known to be needed in the natural habitat.

A plastron has a characteristic sheen, because of the trapped gas. If a piece of *Aphelocheirus* plastron is watched while high pressures are applied to it, it looks normal until the pressure reaches about 4 atm and the pressure difference must be about 3 atm. At higher pressures the sheen fades but it returns in part when the pressure is reduced again and it is believed that the fading is due to the hairs collapsing rather than to the surface pulling away from them.

The hairs can be regarded as slender columns, fixed at one end and free at the other. They presumably collapse by elastic instability, like the column illustrated in Fig. 67c (page 156). What pressure should make them collapse? There are 250 million hairs per square centimetre of plastron so a pressure difference ΔP dyn/cm² would mean a load of $\Delta P/(2\cdot5 \times 10^8)$ dyn on each hair. They are 5 μ (5 × 10⁻⁴ cm) high and have a radius of 0·1 μ. They are probably made of material with a Young's modulus of about 10¹¹ dyn/cm², like locust cuticle (Table 3, page 72). By putting these values in the formula given on page 156, it can be estimated that a pressure difference of 2 atm should make them collapse. This is less than the observed value of about 3 atm, but better agreement could hardly be expected since the estimate was based on rough measurements and a guessed value for Young's modulus. Indeed, the data should perhaps have led to an even lower estimate. We used the formula for a column with a force acting along its long axis, but the force on the bent-over tip of each hair in the plastron must act to one side of the long axis of the hair. We could have used a more complicated formula which would have allowed for this (Warnock and Benham, 1965) but we would then have had to guess the strength of the hairs as well as their modulus.

Crisp (1950) considered another problem concerning the plastron. What will happen if the hairs get slightly bent so that their tips are unevenly spaced? Fig. 85d shows the tips of three hairs. Hair B has been bent to be nearer hair A than hair C. The pressure difference between the water and the air exerts a larger force on the surface between B and C than on the surface between A and B, because its area is greater. The surface tension exerts equal forces on the two sides of B but the surface meets B lower down on the right than on the left, so that the surface tension acts at a steeper angle and has a larger vertical component. In this way the small force exerted by the pressure difference on the surface between A and B and the large force on the surface between B and C can both be balanced. However, the asymmetrical forces exerted by surface tension on B will tend to move it further towards A. If B is stiff enough it will spring back to its original position half-way between A and C but if it is not it will move right up to A. If the hairs of the plastron are not stiff enough the plastron will be unstable because the slightest disturbance of their even spacing will make them clump together, leaving wide gaps. The plastron of *Aphelocheirus* is stable, so its hairs must be sufficiently stiff.

Surface films

The surface tension of water can be reduced greatly by tiny quantities of various materials which form thin films on its surface. Some of these materials, such as the fatty acids, are insoluble. When a water surface is contaminated with them, they spread out to form a film on the surface, a single molecule thick. If there is too little of the material to cover the whole surface, the surface tension is generally not affected much. If it does cover the whole surface it reduces the surface tension because any reduction of the area of the surface squashes the molecules of the film more tightly together.

Some other materials which form surface films on water are more or less soluble. They include the soaps. A soap solution that has been standing for a while has its surface completely covered by a surface film one molecule thick and has a low surface tension (about 25 dyn/cm). If the area of the surface is increased more soap molecules from the solution enter the surface film so that it completely covers the enlarged surface, but this does not happen instantaneously. While the area is increasing the surface film is incomplete and the surface tension higher. This is what makes soap bubbles so stable: if some disturbance stretches the surface of a bubble the surface tension increases and tends to restore it to its original size.

The effects of surface films on surface tension are studied by means of the Langmuir trough and various instruments derived from it, which are described in books on surface tension (for instance, Davies and Rideal, 1961).

Lungs

Fig. 86 shows the results of an experiment with a lung taken from the body of a cat. The continuous line is a graph of the volume of the lung against the pressure in it, as it was inflated with air and deflated again. It forms a marked hysteresis loop, like the loops in graphs of force against length for visco-elastic materials which are stretched and allowed to recoil (Fig. 30, page 70). The loop suggests viscosity in the lung wall but the broken line shows that most of the area of the loop has nothing to do with the properties of the lung wall. This far more slender loop is a graph of volume against pressure when the lung was inflated with a saline solution after the air had been removed from it with a vacuum pump. There is far less hysteresis than when the lung was filled with air, and much less pressure is needed to inflate the lungs. When the lungs are filled with saline there is no air-fluid interface inside them, so the experiment suggests that a good deal of the pressure needed to inflate the lungs is needed to overcome surface tension.

This seems reasonable when one considers the structure of the

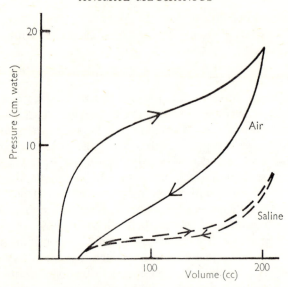

Figure 86. Graphs of pressure against the volume of a cat's lungs as they were inflated with air and with saline solution, and allowed to deflate. (Redrawn from Radford, 1957)

lungs. The air passages in them branch and branch again and finally end in little pockets known as alveoli. The average radius of the alveoli in a half-inflated human lung is about 60 μ, and in a rat lung it is about 20 μ (Clements, Brown and Johnson, 1958). Consider an alveolus of radius 60 μ (0·006 cm) and assume that the fluid wetting it has the same surface tension as blood plasma, about 50 dyn/cm. By equation (20) (page 166) the surface tension will cause a pressure inside the alveolus of 2 × 50/ 0·006 = 1·7 × 10⁴ dyn/cm² or 17 cm water. Smaller alveoli would have bigger pressures.

The difference between the pressure needed to inflate a lung with saline and the pressure needed to inflate it with air is not too large to be explained by surface tension, but it cannot be explained by any constant value of surface tension. If surface tension were constant the pressure it caused in an alveolus would fall, as the alveolus was inflated. The difference between the continuous and broken curves in Fig. 86 would be less when the lung was fully inflated than when it was half inflated. This is not the case. The surface tension must rise, as the lung is inflated. A surface film could make it do this.

If a saline solution is stirred with chopped lung and then filtered, an extract is obtained which is coated with a film of material which apparently coats the internal surface of the intact lung. The surface tension of extracts of this sort rises to a maximum of about 40 dyn/cm as the area of the surface increases and falls to a minimum of about 2 dyn/cm as the area decreases (Fig. 87). There is marked hysteresis. The properties of the surface film are exactly the properties that are needed to explain Fig. 86 (Clements, Brown and Johnson, 1958; Clements, Hustead, Johnson and Gribetz, 1961; Clements, 1962).

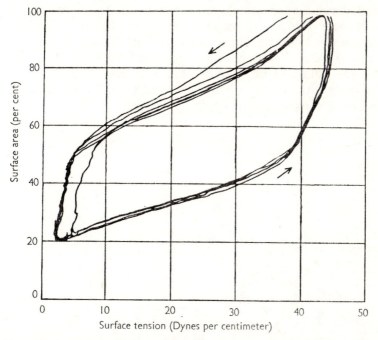

Figure 87. A graph showing how the surface tension of lung extract changes as the surface area is changed. The arrows show the directions of the changes. The loops would run clockwise as in Figs. 30c and 86 if tension had been plotted against area, instead of the other way round. (From Clements, *Surface tension in the lungs*. Copyright © 1962 by Scientific American Inc. All rights reserved)

In certain pathological conditions fluid accumulates in the lungs and a froth is formed. The froth can also be produced artificially. When washed froth is dried a substance with a waxy consistency remains. It is lipoprotein, and it is apparently the material that stabilizes the froth and that forms the natural surface film in the lungs (Pattle, 1965).

The importance of the surface film is shown clearly by the effect of its deficiency in some newborn babies. The condition is called the respiratory distress syndrome. The child has difficulty in breathing and often dies in a few days. If its lungs are examined after it has died, the alveoli are found to be collapsed instead of being filled with air in the usual way. The surface tension of extracts of the lungs shows unusually little hysteresis, and the minimum surface tension is much higher than usual.

If there were no surface film the pressure in a lung would increase as it was deflated and the alveoli got smaller. They would tend to collapse entirely, and a large pressure would be needed to start filling them again for the next breath. This is what happens in the respiratory distress syndrome, although the film is deficient rather than completely absent.

CHAPTER 6

MOTION IN FLUIDS

THIS chapter is concerned with the forces that act when a fluid moves relative to a solid, whether the fluid moves past a barrier or through a pipe, or a body moves through the fluid. The physics of all these situations is dealt with in textbooks on hydrodynamics (for instance, Tietjens, 1957, and Prandtl, 1952) and is introduced in as simple a manner as possible in this chapter.

Reynolds number

The forces that act when a fluid moves relative to a body depend, among other things, on the pattern of flow. One might suppose that flow past spheres, for instance, would always follow the same pattern, but this is not the case. It is only the same if the Reynolds number is the same.

The Reynolds number is a dimensionless quantity. It is $\rho u a/\eta$, where ρ and η are the density and viscosity of the fluid, u is its velocity relative to the body and a is a length. Fluids flowing through tubes flow faster at the centre than near the wall and the mean velocity is used as u. The length a is usually taken as the radius of the tube, when flow through a tube is being considered, and as the length of the body in the direction of flow, when flow past a body is being considered.

The rule that the pattern of flow is the same round bodies of the same shape, if the Reynolds number is the same, does not apply in cases involving a free liquid surface. For instance, Reynolds number is no use for predicting whether the pattern of waves round a ship will match the pattern round a small-scale model.

It is convenient to write Reynolds number ua/v, where $v = \eta/\rho$ and is known as the kinematic viscosity of the fluid. The kinematic viscosities of some fluids are given in Table 4.

TABLE 4

THE KINEMATIC VISCOSITIES OF SOME FLUIDS AT 20°C

(Data from Hodgman, 1965)

	Viscosity (poise)	Density (gm/cm³)	Kinematic viscosity (cm²/s)
Air	$1 \cdot 8 \times 10^{-4}$	$1 \cdot 3 \times 10^{-3}$	$0 \cdot 14$
Water	$0 \cdot 010$	$1 \cdot 00$	$0 \cdot 010$
Glycerin	15	$1 \cdot 26$	12

Pipes, blood vessels and gills

Fluid may flow through a pipe in either of two ways. If the Reynolds number is low, each particle of fluid takes a smooth regular path which is straight if the pipe is straight. This is laminar flow. If the Reynolds number is high the particles swirl about, taking irregular paths. This is turbulent flow. As the velocity of flow in a pipe is increased a critical velocity is reached at which the flow changes suddenly from laminar to turbulent. This usually happens when the Reynolds number is about 1,000.

It may be helpful to give some examples of cases where the Reynolds number would be 1,000. Water flowing at 10 cm/s in a pipe of radius 1 cm or at 100 cm/s in a pipe of radius $0 \cdot 1$ cm, or air travelling at 140 cm/s in a pipe of radius 1 cm, would have a Reynolds number of 1,000.

Poiseuille's equation gives the pressure needed to drive a fluid through a pipe, when flow is laminar. Consider a pipe of length l and radius r. A fluid of viscosity η is driven through it at a rate (volume per unit time) ω. The pressure difference between the ends of the pipe must be ΔP where

$$\Delta P = 8\eta l\omega/\pi r^4 \qquad (26)$$

The relationship is more complicated when flow is turbulent (Tietjens, 1957).

The work needed to drive a fluid through a pipe is the product of the volume which flows and the pressure difference. The pressure difference is proportional to the rate of flow, ω, when flow is laminar (and about to $\omega^{1 \cdot 8}$ when flow is turbulent). Therefore more work is needed to drive the fluid rapidly through the pipe than to drive it slowly and more work is needed to drive a given quantity of fluid through the pipe in a given time if it is driven through in a series of spurts than if it is driven steadily by a constant pressure difference.

If arteries had rigid walls the blood would travel through the capillaries in a series of spurts. It would move as long as the heart was contracting, and then remain stationary until it contracted again. This does not happen. The arteries swell as the heart contracts, storing both blood and elastic energy. The elastic recoil drives the stored blood on through the smaller blood vessels in the interval between contractions. In this way the velocity of flow through the arterioles and capillaries is kept much steadier than it would otherwise be. Less energy is needed to drive the blood round the body than if the arteries had rigid walls.

The elasticity of arteries is largely due to the elastin in their walls. They also contain collagen which is much less extensible than elastin, but the collagen fibres seem to be arranged loosely so that they do not become tight until the artery is distended. The part of the aorta that runs through the thorax is particularly distensible. In dogs, its diameter increases by 10% at each heart beat, while the rest of the aorta and the femoral artery swell by only 5% and 4%. It contains about 30% elastin, while the rest of the aorta and the other arteries contain only about 15% (Harkness, Harkness and McDonald, 1957; these are percentages of the dry weight). Because it contains a lot of elastin that can be stretched, the thoracic part of the aorta can store a good deal of elastic energy and is particularly effective in smoothing the flow of blood.

When fish pump water through their gills it passes through the channels between the gill lamellae. These channels are, in effect, a battery of very fine pipes, and more energy would be used driving the water through if it were driven in spurts, than if it were driven steadily. The water is pumped by movements of the mouth cavity and of the gill cavities, and these movements are generally coordinated so as to keep the flow of water reasonably steady for most of each cycle of breathing movements (Hughes and Shelton, 1962; Alexander, 1967b).

Fluids flow faster in the centre of a pipe than near the walls. This is not so at the mouth of a pipe where fluid enters it from a large tank, but as the fluid flows along the pipe the fluid near the wall is slowed down by viscosity until there is a gradual increase of velocity from zero at the wall to a maximum at the centre (Tietjens, 1957).

Fig. 88*a* shows how water travels in a pipe when it is divided into lengths trapped between air bubbles. The arrows in the water represent the velocity of the water *relative to the bubbles*. The bubbles and the water are all travelling to the right relative to the pipe but the water near the wall of the pipe is slowed down by viscosity. It travels more slowly than the bubbles and therefore its velocity relative to the bubbles is to the left. The water at the centre of the tube travels faster than the bubbles so the water circulates as the arrows indicate as it moves towards the right. This sort of flow is called bolus flow (Prothero and Burton, 1961). It presumably occurs in blood capillaries because the red corpuscles are larger in diameter than the capillaries and travel along them, one behind the other, squashed out of shape (Fig. 88*b*; Guest, Bond, Cooper and Derrick, 1963). The circulation of the plasma as it travels along between the corpuscles must aid the diffusion of substances between the blood and the tissues. For instance, oxygen that diffuses out of the centre of a corpuscle will be carried round in this circulation to the wall of the capillary.

Bernoulli's theorem

Bernoulli's theorem describes steady flow along a streamline. It is necessary to explain both what steady flow is, and what a streamline is. Flow is called steady if the velocity of flow is

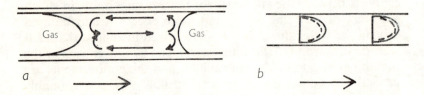

Figure 88. (*a*) A column of water flowing from left to right along a tube between two bubbles of gas. The arrows in the water show movement relative to the bubbles. (*b*) A diagram showing how red blood corpuscles are deformed in capillaries

constant at every point in space, even if the velocity of the fluid changes as it goes along. Flow in a pipe is steady if the velocity in each part of the pipe is constant, even if the pipe gets wider and narrower and the fluid travels slower and faster as it goes along. A streamline is a curve whose direction is everywhere the same as the direction of flow of the fluid. If flow is steady a molecule starting at the upstream end of a streamline will follow the whole course of the streamline.

Bernoulli's theorem relates the velocity u, height h and pressure P of a fluid at different points along the same streamline. It applies only when flow is steady. For each point on the streamline

$$u^2/2 + gh + P/\rho = k \qquad (27)$$

where ρ is the density of the fluid and k is a constant which is the same for all points on the same streamline but is not necessarily the same for other streamlines. Notice that if the fluid is flowing horizontally so that h is constant, an increase in velocity will mean a decrease in pressure. We will see in the next section of this chapter how this effect can be used to measure the velocity of a fluid. It is also used in laboratory filter pumps. The water leaves a filter pump at atmospheric pressure and at a relatively low velocity, but there is a constriction inside the pump where it travels much faster and has a correspondingly lower pressure. The apparatus that is to be evacuated is connected to this region of low pressure.

Bernoulli's theorem should not be applied to flow very close to a solid body.

Wind tunnels and water tunnels

It is often difficult to measure the forces which act on a body as it moves through a fluid. It is often more convenient to measure the forces with the body stationary and the fluid moving. Wind tunnels and water tunnels are used in experiments of this sort. They have to be very carefully designed, if the forces are to be the same as in the natural situation with the body moving. In the natural situation the fluid is typically still, except immediately round the body. The bulk of the fluid therefore has a uniform velocity relative to the body. The apparatus should be designed so that the fluid approaching the body has a uniform velocity and is free from eddies.

These are difficult requirements. When a fluid flows in a tube its velocity is not uniform but is less near the walls of the tube than near the centre. If the tube is reasonably wide and the flow reasonably fast, the flow is turbulent.

Fig. 89 shows one type of wind tunnel, in which these difficulties are largely overcome (Prandtl, 1952). Air is driven by the fan into the wide chamber where its velocity is low. The screen of wire gauze helps to equalize the velocity, across the cross-section of the chamber. The honeycomb is a grid of metal plates like the grids that separate the ice cubes in some refrigerators, and it ensures that there is no large-scale swirling around in the chamber, but that the air travels along it in reasonably straight lines. Thus, the wind in the chamber is reasonably uniform and reasonably straight. Such irregularity as there is in the wide chamber is swamped by the large, uniform increase of velocity that occurs when the air passes through the much narrower nozzle. The manometer indicates the wind speed.

Some wind tunnels are totally enclosed, with the specimen in a pipe which is an extension of the nozzle and leads back to the fan. Water tunnels are built on the same principles as wind tunnels and are, naturally, enclosed.

There are various techniques for making visible the patterns

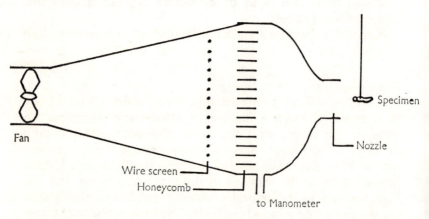

Figure 89. An open-jet wind-tunnel of the type used by Weis-Fogh (1956a)

of flow round specimens in wind tunnels and water tunnels (Tietjens, 1957). It is sometimes useful to be able to see where there are eddies.

Drag

When a body moves through a fluid a force acts backwards on it, resisting its motion. This force is known as drag. Its amount depends on the fluid and on the size, shape and speed of the body.

Experiments in wind tunnels have shown that the drag on a body or on different bodies of the same shape is given by the equation

$$D = \tfrac{1}{2}\rho u^2 A C_\mathrm{D} \qquad\qquad (28)$$

where D is the drag, ρ is the density of the fluid, u is the velocity of the fluid relative to the body, A is a characteristic area (this will be explained) and C_D is a quantity known as the drag coefficient which depends on the shape of the body and on the Reynolds number. If the drag coefficient had been defined differently the factor $\tfrac{1}{2}$ could have been left out of the equation: it is included because of the term $u^2/2$ in Bernoulli's equation (equation 27, page 213).

We have still to define the area A. There is unfortunately no single definition which is convenient for all shapes of body. The following areas are used:

(i) the frontal area, which is the projected area of the body in the direction of flow ("projected area" was explained on page 165). In the case of a cylinder of height h and radius r this would be πr^2 if flow was parallel to the axis of the cylinder and $2rh$ if it was at right angles to the axis;

(ii) the greatest projected area, which is the projected area in the direction in which it is greatest. This area is used for aerofoils and has the advantage over frontal area of remaining constant as the aerofoil is tilted;

(iii) the total surface area. Note that in the case of a plate this would be the total area of both sides of the plate.

It is important to state which of these areas has been used to calculate a drag coefficient, whenever doubt is possible.

Fig. 90 is a graph of drag coefficient against Reynolds

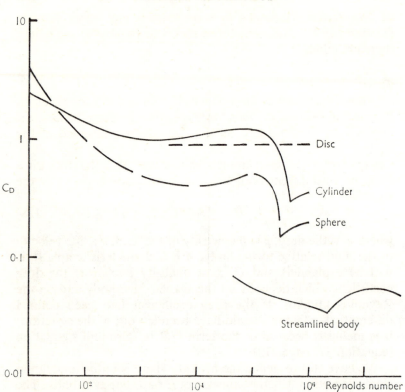

Figure 90. Graphs of drag coefficient against Reynolds number for a disc moving at right angles to its plane, a long cylinder moving at right angles to its axis, a sphere, and a streamlined body moving along its axis. (Data from Tietjens, 1957, and Landolt and Börnstein, 1955)

number, for bodies of various shapes. A streamlined body is more or less rounded at the front and tapers gradually to a point at the back, like a torpedo without fins (Fig. 91*a*, *b*). All the coefficients have been calculated from the frontal area. Reynolds number has been calculated in the usual way from the length measured in the direction of flow, for all the bodies except the disc. The diameter of the disc has been used although it is at right angles to the direction of flow.

At low Reynolds numbers, below about 100, drag coefficients change rapidly with Reynolds number. Reynolds number is

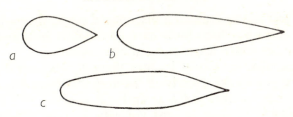

Figure 91. (*a*), (*b*) Streamlined bodies of fineness ratio 2 and 4·5, respectively. (*c*) An aerofoil section which is described in the text. Its thickness is exaggerated. The front ends of (*a*), (*b*) and (*c*) are all towards the left

about 100 for a body 1 cm long moving through water at 1 cm/s, for a body 0·1 cm long moving through water at 10 cm/s or for a body 1 cm long moving through air at 14 cm/s. More will be said later about drag at extremely low Reynolds numbers.

The drag coefficient is nearly constant for a body of given shape moving in a given direction, at Reynolds numbers between about 10^3 and 10^5. It has a minimum value at a critical Reynolds number which is usually between 2×10^5 and 2×10^6. In the cases of spheres and cylinders it drops so abruptly to the minimum that an increase in velocity actually results in a fall in drag. To reach its critical Reynolds number a body must be by zoological standards large or fast or both. The Reynolds number would be 10^6 for a body 10 cm long moving through water at 10 m/s or through air at 140 m/s, or for a body 100 cm long moving through water at 1 m/s or through air at 14 m/s.

There are wide differences between the drag coefficients at moderate Reynolds numbers, of bodies of different shapes. A disc set at right angles to the flow gives a drag coefficient of about 1·1 over a wide range of Reynolds numbers, and parachute shapes give even higher coefficients. The lowest drag coefficients, about 0·05, are given by streamlined bodies.

It will help us to understand why shape affects drag so much, if we distinguish two types of drag. These are pressure drag and friction drag. When a body travels through a fluid it leaves a wake in which the fluid is swirling around. This fluid has been given kinetic energy, and this energy must have come from work done in driving the body through the fluid against drag. The part of the drag accounted for in this way is called pressure drag. Most of the drag on a disc moved perpendicularly to its plane is pressure drag. So is most of the drag on spheres and other bodies which leave very disturbed wakes.

A thin plate moved in its own plane leaves very little disturbance in its wake. Nearly all the drag is friction drag, which corresponds to work done against the viscosity of the fluid. Work has to be done against viscosity, because the fluid immediately in contact with the body is carried along with it and there is a velocity gradient in a thin layer of fluid (known as the boundary layer) which covers it. Because there is so little pressure drag, the total drag on a thin disc is far less when it moves in its plane than when it moves in a direction perpendicular to its plane. At a Reynolds number of 10^4, for instance, it is only 1/40 as much. The gradual taper at the back of a streamlined body encourages the fluid to flow in smoothly behind the body so that there is not much disturbance in the wake. The drag on a well streamlined body is mainly friction drag and so is not much more than the drag on a flat plate of equal total area, moving in its own plane.

The drag on a streamlined body depends on the ratio of the length to the maximum diameter, which is known as the fineness ratio. Increasing the fineness ratio increases the surface area and the friction drag, but it decreases the pressure drag. If bodies of equal frontal area are compared, the drag is least when the fineness ratio is 2 (Fig. 91a). If bodies of equal volume are compared, it is least when the fineness ratio is about $4 \cdot 5$ (Fig. 91b) but it is only about 10% higher when the fineness ratio is 3 or 7. Fineness ratios around $4 \cdot 5$ were naturally preferred for airship hulls. The pressure drag on a well streamlined body of fineness ratio $4 \cdot 5$ is only about 20% of the total drag (Zahm, Smith and Louden, 1928. The figures given in this paragraph were obtained in experiments at Reynolds numbers around 10^7, and do not necessarily apply to lower Reynolds numbers.)

The kinks in the graphs of drag coefficient, at Reynolds numbers of the order of 10^6, are due to a change in the boundary layer. At lower Reynolds numbers the flow in the boundary layer is laminar and at higher ones it is turbulent. In the region of the kink, flow is laminar over the front part of the body and turbulent towards the rear. The phenomenon is the same as the change from laminar to turbulent flow that occurs at a Reynolds number of about 10^3 in pipes. The difference between the critical Reynolds numbers (10^6 for flow over bodies; 10^3 for flow in pipes) is due simply to the two sorts of Reynolds number being calculated in different ways. A Reynolds number calculated from the radius of a pipe and one calculated from the length of a body are essentially different quantities. While flow is laminar the friction drag on a body is about $\frac{1}{2}\rho u^2 S$ ($1 \cdot 33\ R^{-0.5}$), where ρ is the density of the fluid, u the velocity, S the total surface area and R Reynolds number. When flow becomes turbulent, the friction drag becomes $\frac{1}{2}\rho u^2 S(0 \cdot 074\ R^{-0.2})$. This means that the onset of turbulence increases friction drag. However, a turbulent boundary layer tends to make the wake smaller and so to reduce pressure drag. This is why the drag on some bodies actually falls as the velocity increases, at the stage when turbulence appears.

There is a range of Reynolds numbers in which the drag coefficient of a streamlined body can be reduced, by preventing part or all of the boundary layer from becoming turbulent. It has been shown in experiments with aerofoils that the drag can be kept below the value for a fully turbulent boundary layer up to abnormally high Reynolds numbers, if the thickest part of the aerofoil is set far back (Fig. 91c). The boundary layer tends to remain laminar as far back as the thickest part, but if the thickest part is set too far back the advantage is cancelled out by an increase in pressure drag. Similar effects could presumably be obtained with torpedo-shaped bodies, by setting the thickest part well back. It may also be possible to maintain laminar flow beyond the usual critical Reynolds number if the body has a coating of spongy material (see page 221). Gadd (1966) has found that certain substances reduce turbulence in water markedly when they are present in tiny quantities, but there do not seem to have been any suggestions so far that effects of this sort may be important in zoology.

Roughness has little effect on drag unless the irregularities of the surface are big enough to protrude from the boundary layer. Since the boundary layer gets thinner as Reynolds number increases, irregularities that are unimportant at low speeds may have a serious effect at high ones. On a body of length l, projections of height $10^{-3} l$ become important at a Reynolds number of 10^5, ones of height $10^{-4} l$ at 10^6, and so on (Prandtl, 1952).

We must now consider drag at lower Reynolds numbers than we have considered so far. A body 1 mm long moving through water at 1 mm/s would have a Reynolds number of 1. When the Reynolds number is less than 1, the drag on a body of length a travelling at velocity u through a fluid of viscosity η is given by the equation

$$D = k\eta a u \qquad (29)$$

k is a constant which depends on the shape of the body: for a sphere of diameter a, it is 3π. Notice how different this is from equation (28); neither the density of the fluid nor the area of the body appears in it, and the velocity is not squared. Stoke's law follows from equation (29): a sphere of radius r and density ρ sinking in a fluid of density ρ' and viscosity η will accelerate until its velocity is u, where

$$u = 2r^2 g(\rho - \rho')/9\eta \qquad (30)$$

The next three sections of this chapter deal with zoological problems concerning drag at very different Reynolds numbers.

Drag on dolphins

Dolphins are large animals which swim fast and so have large Reynolds numbers. *Tursiops gilli* is a Pacific member of the same genus as the British bottle-nosed dolphin. Lang and Norris (1966) found that it could get up to 830 cm/s (about 16 knots) in a short burst of speed, and could keep going at 610 cm/s (about 12 knots) for nearly a minute. Their specimen was 191 cm long so the Reynolds number at the higher speed was $830 \times 191/0 \cdot 01 = 1 \cdot 6 \times 10^7$.

The body of *Tursiops* looks well streamlined, apart from a concavity on the top of the snout which is the reason for its being called "bottle-nosed". The body is torpedo-shaped with a roughly circular cross-section and tapers gently at the posterior end. Its fineness ratio is about 5, close to the ideal ratio of $4 \cdot 5$. The only projections from the body are the flippers, dorsal fin and flukes, which all have streamlined cross-sections (page 236). The skin is very smooth and has no hair. Everything points to the drag being low. Steven (1950) watched dolphins and seals swimming at night among plankton which phosphoresced wherever the water was disturbed. The seals left a bright phosphorescent wake, but the wakes of dolphins were far less conspicuous. This seems to confirm what one would suppose from their shape, that dolphins leave relatively little disturbance in their wake and so suffer little pressure drag.

We will estimate the drag on *Tursiops*, swimming at 830 cm/s, and the power output of its muscles. The frontal area of the specimen, 191 cm long, must have been about 1,100 cm². The drag coefficients of streamlined bodies at Reynolds numbers around $1 \cdot 6 \times 10^7$ are about $0 \cdot 055$ (Fig. 90, page 216). Putting these values in equation (28) we estimate that the drag on the dolphin must have been about $\frac{1}{2}(830)^2 \times 1,100 \times 0 \cdot 055 = 2 \cdot 0 \times 10^7$ dyn. The power is the drag multiplied by the speed and is $830 \times 2 \cdot 0 \times 10^7$ erg/s or 1,660 W. This is the power exerted against drag on the body. The total power required of the muscles must be more than this, for the swimming action cannot be 100% efficient (see page 277). It could hardly be much less than 2,000 W. This *Tursiops* weighed 89 kg, of which about 15 kg must have been swimming muscle. We must estimate the power output of the muscles as about 130 W/kg. This is three times the maximum power that people can get from their muscles, on a bicycle ergometer (page 24).

Do dolphin muscles really produce as much power as this? The power output of dolphins has interested many zoologists since Gray (1936) first estimated it. Various zoologists have suggested that the flow in the boundary layer may somehow be kept largely laminar, in spite of the high Reynolds number.

The outline of a dolphin's body resembles the outline of the aerofoil section shown in Fig. 91c. The thickest part lies well back, about 55% of the body length from the snout. The flow might remain laminar as far back as this. Aerofoils with this sort of section get away with only 65% of the friction drag they would suffer if flow were completely turbulent, at a Reynolds number of $1 \cdot 6 \times 10^7$ (Hertel, 1966).

Dolphins have a thin spongy layer in the outer part of their skin. Local fluctuations of pressure such as occur in a turbulent boundary layer distort the spongy material, but the distortion is resisted by the viscosity of the fluid which permeates the sponge. The skin might keep the boundary layer laminar by damping out pressure fluctuations. Kramer (1960, 1965), who suggested this, made a model with artificial coatings imitating dolphin skin. The most successful coating reduced the drag at a Reynolds number of $1 \cdot 6 \times 10^7$ to about 40% of the normal value. Other investigators seem to have been unable to repeat these remarkable results (Gadd, 1963).

Lang and Pryor (1966) approached the problem of dolphin swimming in a different way. They filmed a specimen of the dolphin *Stenella*, coasting along with its body straight after a burst of fast swimming. It was of course slowed down by the drag, which was high at first and less as it slowed down. Lang and Pryor measured the deceleration at a series of speeds, and used Newton's second law of motion (equation 1, page 1) to calculate the drag at each speed. They concluded from the values they obtained that the boundary layer was fully turbulent.

Drag on water beetles

The water beetle *Acilius* is about $1 \cdot 7$ cm long and can just get up to a speed of 50 cm/s as it swims under water. Its Reynolds number at this speed is $1 \cdot 7 \times 50/0 \cdot 01 = 8,500$. *Dytiscus* is bigger and may be able to swim faster, but the Reynolds numbers at which both beetles swim are plainly in the range where the drag coefficient is more or less independent of Reynolds number (Fig. 90, page 216). Flow in the boundary layer must be laminar and no special precautions are needed to keep it so.

The bodies of the beetles do not look very well streamlined. Nachtigall (1960) and Nachtigall and Bilo (1965) determined their drag coefficients, in the appropriate range of Reynolds numbers. Some of the measurements were made in a water tunnel, with the water flowing at the natural swimming speed. Others were made in a wind tunnel. As the kinematic viscosity of air is 14 times the kinematic viscosity of water, the air had to be driven 14 times as fast to obtain the same Reynolds number. The drag coefficient of *Acilius*, based on frontal area, was found to be $0 \cdot 23$, and that of *Dytiscus* $0 \cdot 34$. A well streamlined body at the same Reynolds number would probably have a drag coefficient of less than $0 \cdot 1$ (Fig. 90, page 216). Plainly the beetles are not well streamlined and suffer a lot of pressure drag. Their method of swimming is described later in this chapter (page 225).

Theoretical efficiency and hydraulic efficiency

When an aeroplane or a boat is driven by a propeller, the forward thrust that is needed to overcome drag is obtained by driving the fluid backwards. If a mass m of fluid is accelerated every second to a velocity u, momentum is given to the fluid at a rate mu per second and by Newton's second law of motion (page 1) the thrust is mu. The power needed for accelerating the fluid is the rate at which kinetic energy is given to it, $\frac{1}{2}mu^2$ per second. Obviously a given thrust can be obtained more economically, if m is large and u is small than if u is large and m is small. It is more economical to propel a vehicle by accelerating large quantities of fluid to a small velocity, than smaller quantities to a higher velocity.

The total energy used in driving the propeller can be divided into three parts. The first, for which we will use the symbol E_1 is used in doing work against drag on the vehicle. The second, E_2, is used in giving backward momentum to the fluid to drive the vehicle forwards. The third, E_3, is energy lost in various other ways, such as in forming eddies in the wake or in overcoming drag on the blades of the propeller (Prandtl, 1952).

It will be convenient to express efficiencies as fractions, rather than as percentages. The efficiency of the propeller is the energy used overcoming drag on the vehicle divided by the total energy or $E_1/(E_1 + E_2 + E_3)$. This is the total efficiency and it is convenient to define as well a theoretical efficiency $E_1/(E_1 + E_2)$ and a hydraulic efficiency $(E_1 + E_2)/(E_1 + E_2 + E_3)$. Notice that these are defined in such a way that the total efficiency is (theoretical efficiency) $\times$ (hydraulic efficiency). The theoretical efficiency depends on such things as the size of the propeller relative to the vehicle. It follows from what was said earlier that it is highest when relatively large quantities of fluid are being accelerated to a small velocity. A high theoretical efficiency demands a large propeller. The hydraulic efficiency depends on such things as the shape of the propeller, and is usually about 0·85 (85%) for well-designed propellers.

Animals do not have propellers, but the concepts of theoretical and hydraulic efficiency are still useful in zoology. We will use them in discussing the flight of birds.

Rowing

Rowing consists of power strokes in which the oars move backwards relative to the boat, alternating with recovery strokes in which they move forward again. Fig. 92 represents a rowing boat which we will suppose moves with constant velocity U although a real boat would accelerate in the power stroke and slow down in the recovery stroke. Drag D acts on it. The blades of the oars move forward and back with velocity u relative to the boat, so that they have a backward velocity $(u - U)$ relative to the water in the power stroke and a forward velocity $(u + U)$ relative to the water in the recovery stroke. Drag d acts on each oar in the power stroke and d' in the recovery stroke. The power needed to overcome the drag on the boat is DU. In the power stroke, power $2d(u - U)$ is used overcoming drag on the two oars and the total power is $[2d(u - U) + DU]$. In the recovery stroke, power $2d'(u + U)$ is used overcoming drag on

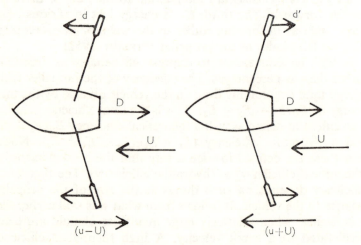

Figure 92. Diagrams illustrating the discussion of rowing

the oars and the total power is $[2d'(u + U) + DU]$. The mean power consumption is $[d(u - U) + d'(u + U) + DU]$ and the efficiency is $DU/[d(u - U) + d'(u + U) + DU]$. A forward force $2d$ acts on the oars in the power stroke and a backward force $2d'$ acts on them in the recovery stroke so the mean force on the oars is $(d - d')$, forwards. This must balance the drag on the boat so $D = (d - d')$. By substituting this in our expression for efficiency, we find that the efficiency is (U/u) $[(d - d')/(d + d')]$. The first factor in this expression, (U/u), depends on the relative sizes and drag coefficients of the boat and the oars, and may be compared with the theoretical efficiency of a propeller. The efficiency will actually be less than the expression, because we have ignored some ways in which energy is lost. For instance, oar blades do not move straight but swing in an arc and we have not allowed for the work done unprofitably, pushing water sideways.

For the efficiency to be high, the speed of the oars, u, must not be much greater than the speed of the boat, U. Oars should be designed to give large drag at low velocity. Fast rowing boats are streamlined but their oars have large flat blades which are held perpendicular to their path through the water to get as much drag as possible. d' must be kept small. Oarsmen remove their oars from the water for the recovery stroke, because drag is much less in air than at the same speed in water. They feather their oars (i.e. they turn them so that the blades are horizontal) to reduce d' still further.

Swimming of water beetles

Hughes (1958) and Nachtigall (1960) used cinematograph films to study the swimming movements of the water beetles *Dytiscus* and *Acilius*, which use their middle and hind legs to row themselves along under water. The hind legs are larger than the middle ones and only they are shown in Fig. 93, which shows them at four stages in the power stroke (1 to 4) and four in the recovery stroke (5 to 8). The left and right legs move together.

One of the hind legs is conveniently displayed in the side view, in position 4. It is quite unlike the legs of ordinary beetles. The femur, tibia and tarsus are elliptical in cross-section, instead of

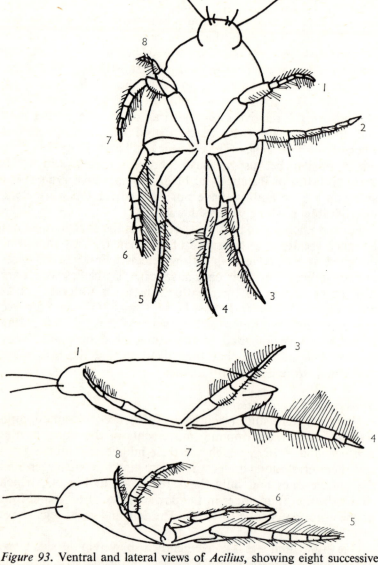

Figure 93. Ventral and lateral views of *Acilius*, showing eight successive positions of the hind leg in swimming movements. The other legs are not shown. (Redrawn from Nachtigall, 1960)

circular (for the names of the parts of insect legs, see Fig. 12a, page 00). The tibia and tarsus have long hairs attached to their narrow edges. The tarsus is very long and the tibia relatively short. The coxa is fixed rigidly to the body. The joints between the coxa and the trochanter and between the femur and the tibia are hinge joints, each allowing rotation about a vertical axis. The joint between the tibia and the tarsus is a ball and socket joint.

Because of this joint the tarsus can be feathered like an oar. Its broad faces are vertical in the power stroke but horizontal in the recovery stroke. Films of beetles which had one face of the tarsus painted white showed that the face which is anterior in the power stroke becomes dorsal in the recovery stroke. Feathering is automatic, because the hairs along the dorsal edge of the tarsus are much longer than the hairs along the ventral edge. This is shown clearly in the side view (Fig. 93) in position 4. Consider the left leg, as seen in the figure. In the power stroke, drag on the long dorsal hairs tends to rotate the tarsus anti-clockwise but the structure of the joint prevents rotation beyond the position with the broad faces vertical. In the recovery stroke, drag on the long hairs rotates the tarsus clockwise so that these hairs point posteriorly.

The automatic feathering mechanism has another refinement, for the hairs are jointed to the leg in such a way that they spread in the power stroke and trail behind in the recovery stroke.

Nachtigall observed that during the power stroke when the beetle is swimming steadily the joint between the tibia and the tarsus of *Acilius* remains stationary relative to the water. This is a rather surprising observation, for it means that the hairy tibia moves forward in the power stroke and hinders rowing rather than helps it. However, when the beetle is accelerating from rest the whole leg moves backward relative to the water, and the hairs on the tibia must then be helpful. We can use Nachtigall's observation to estimate the factor U/u in the efficiency. Drag acts on all parts of the tarsus, and different parts move at different speeds, but by inspection of the tarsus it seems likely that the resultant drag will act about half way along it, that is about 79% of the length of the leg from the thorax. The joint which remains stationary is 57% of the way along the leg. Hence $U/u = 57/79 \simeq 0.7$.

Nachtigall measured the drag on an *Acilius* leg in a water tunnel. For one set of measurements he fixed the leg so that the current struck it from behind and spread it in the position of the power stroke. For another, he reversed it so that it adopted the feathered position of the recovery stroke. From the results, he calculated a factor in the efficiency, to take account of energy used in the recovery stroke. This factor was equivalent to the factor $[(d - d')/(d + d')]$ on page 225, and came out as about 0.8. He calculated another factor to take account of energy wasted pushing water sideways, as the leg swung in its arc. This also came out as about 0.8. Hence the total efficiency of the rowing mechanism of *Acilius* can be estimated as $0.7 \times 0.8 \times 0.8 = 0.45$.

Aerofoils and hydrofoils

Drag is not the only hydrodynamic force that acts on bodies moving through fluid and bodies with fluid moving past them. By definition, drag acts in the same direction as the velocity of the fluid relative to the body. When a symmetrical body moves along its axis of symmetry, the hydrodynamic force on it acts directly backwards and it is all drag. When a symmetrical body moves at an angle or when an asymmetrical body moves in any but certain particular directions, the hydrodynamic force acts at an angle to its path. It can be resolved into a component acting directly backwards, which is drag, and a component at right angles to the drag. This second component is called lift because the most familiar example of a force of this sort is the upward force which acts on the wings of aeroplanes and keeps them in the air. In spite of its name, lift does not necessarily act upwards.

Structures like aeroplane wings which are designed to produce lift when they move through air are called aerofoils. Structures designed to produce lift in water are called hydrofoils. Fig. 94*a* represents a section through an aerofoil or hydrofoil and shows the directions of lift and drag. The amounts of lift and drag depend on the angle of attack, which is the angle of the aerofoil or hydrofoil relative to the flow of air or fluid. The angle α is the

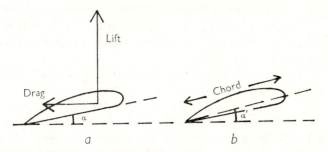

Figure 94. Sections of aerofoils which are moving horizontally to the left, relative to the air. The meanings of some terms are illustrated. (*a*) and (*b*) illustrate alternative definitions of angle of attack

angle of attack according to one definition but the slightly larger angle α', shown in Fig. 94*b*, is the angle of attack according to another.

Drag coefficient is defined by equation (28) (page 215). Lift coefficient is defined by a very similar equation which gives the lift L on an aerofoil or hydrofoil of area A moving with velocity u through fluid of density ρ. The area taken as A is the greatest projected area (see page 215). The equation is

$$L = \tfrac{1}{2}\rho u^2 A C_L \qquad (31)$$

C_L is the lift coefficient. Its value for a particular aerofoil depends on both the Reynolds number and the angle of attack. Aerofoils of the same shape at the same angle of attack have the same lift coefficient, provided the Reynolds numbers are the same. It does not matter whether the sizes of the aerofoils or the velocities or the fluids are different, provided only that the Reynolds numbers are the same. The characteristic length used in defining Reynolds number for aerofoils is the chord length (Fig. 94*b*).

The lift coefficient of an aerofoil at a given Reynolds number is generally roughly proportional to the angle of attack, up to a certain limit. The limit is due to the phenomenon called stalling, and the angle of attack at which stalling occurs is known as the stalling angle. As the angle of attack is increased above the stalling angle, the lift coefficient falls. Stalling involves a marked change in the pattern of flow of fluid over the aerofoil. When the angle of attack is less than the stalling angle, the flow is mostly parallel to the surface of the aerofoil. When it is greater than the stalling angle, large eddies form.

Lift is due to fluid flowing faster over one surface of the aerofoil than over the other. In the case of an aerofoil arranged as in Fig. 94, the fluid flows faster over the upper surface than over the lower one for reasons which do not concern us. The faster fluid above the aerofoil has a lower pressure than the slower fluid below it, as Bernoulli's equation requires (equation 27, page 213), so an upward force acts on the aerofoil. This is the lift.

Fluid flows round the lateral ends of an aerofoil, from the region of high pressure below to the region of low pressure above (Fig. 95). It gains a certain amount of kinetic energy as it does so and it keeps this kinetic energy after the aerofoil has passed. Consequently, the air behind the left wing tip of an aeroplane whirls in one direction, and the air behind the right wing tip whirls in the other. This explains a part of the drag on the aerofoil, which is known as the induced drag. We have already seen that pressure drag is due to kinetic energy lost in the wake (page 217).

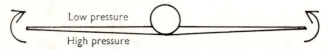

Figure 95. Diagrammatic front view of an aeroplane, referred to in the explanation of induced drag

The amount of induced drag depends on the pressure difference between the upper and lower surfaces of the aerofoil (i.e. on the lift) and on the proportions of the aerofoil. More fluid flows round the end of the aerofoil when the pressure difference is high than when it is low, and more flows round when the chord is long than when it is short. There is an equation which relates the drag coefficient to the lift coefficient and to the aspect ratio, which is the span of the wing (the distance from one wing tip to the other) divided by the average chord. The drag coefficient C_D of an aerofoil of aspect ratio R

at an angle of attack at which its lift coefficient is C_L is given by
the equation

$$C_D = C_{DO} + kC_L^2/\pi R \qquad (32)$$

C_{DO} is the drag coefficient at the angle of attack at which there
is no lift. Its value depends on the shape of the aerofoil as seen in
section (as in Fig. 94). Since the shape of the section is called the
profile of the aerofoil, the part of the drag corresponding to
C_{DO} is called profile drag. The rest of the drag is induced drag.
The value of the constant k in equation (32) depends on how the
aerofoil tapers towards its tips. It cannot be less than 1 but can
be taken as about equal to 1 for most practical purposes.

The properties of an aerofoil can be investigated by measuring the lift and drag on it, arranged at various angles of attack
in a wind tunnel. If it is not convenient to use the actual

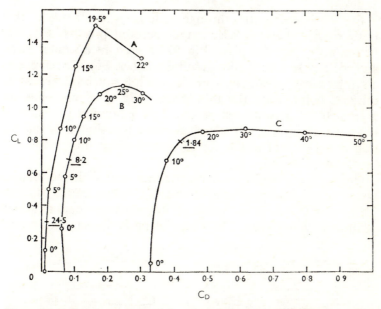

Figure 96. Polar diagrams for (*A*) an aerofoil with a streamlined profile at
a Reynolds number of about 5×10^6, (*B*) a locust hindwing at a Reynolds
number of about 4,000 and (*C*) a *Drosophila* wing at a Reynolds number
of about 200. The maximum lift/drag ratio is shown (underlined) at the
appropriate point on each curve. (From Vogel, 1967)

aerofoil a model may be used, provided the Reynolds number is kept right. The results of experiments of this sort are usually displayed as polar diagrams, which are graphs of lift coefficient against drag coefficient with the corresponding angle of attack written beside each point. Curve A in Fig. 96 is a polar diagram for a model of an aeroplane wing. This aerofoil had a small lift coefficient when the angle of attack was zero, and the lift coefficient increased roughly in proportion to the angle of attack until it reached a maximum value of about 1·5 at 19·5°. The drag coefficient increased as the lift coefficient increased, according to equation (32). 19·5° was the stalling angle, and at higher angles of attack the lift coefficient diminished although the drag coefficient continued to rise. The lift on the aerofoil was 24·5 times the drag when the angle of attack was 2·5° (this point on the curve is marked), and less at other angles of attack.

Fig. 97 shows sections through various types of aerofoil: that is to say, it shows a selection of aerofoil profiles. Fig. 97a represents a thin, flat plate. Fig. 97b represents a similar plate, arched. This sort of arching is called camber. Fig. 97c represents a symmetrical, streamlined profile which is much thicker, at its thickest point, than the thin plates. Fig. 97d represents an asymmetrical, streamlined profile. It is a cambered version of the previous profile, as the arched centre line shows. This is the traditional type of profile for aeroplane wings. Fig. 97e, f will be discussed later.

Ordinary textbooks on aerodynamics deal with the relative merits of different profiles at Reynolds numbers of the order of 10^6 or more. This is the range of Reynolds numbers at which aeroplane wings work. The wings of birds, insects and model

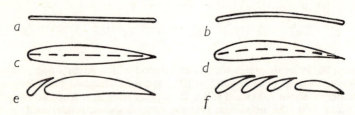

Figure 97. Aerofoil sections which are discussed in the text

aeroplanes are smaller and travel more slowly, and so work at lower Reynolds numbers. Schmitz (1960), who was interested in the design of model aeroplanes, measured the lift and drag on aerofoils of various profiles, at Reynolds numbers down to 2×10^4. Some of his results are shown in Fig. 98, which consists of polar diagrams for two different Reynolds numbers for a flat plate, a cambered plate and a conventional aerofoil. The profiles of these three were very like the profiles shown in Fig. 97a, b and d, respectively.

There are two ways of judging these profiles. For some purposes, the best profile is the one which will give the largest lift coefficient. For others, the best will be the one which gives the highest ratio of lift to drag. It is not necessary for us to choose between these criteria, because they lead us to similar conclusions in the comparison we are making. At a Reynolds number of 168,000, the conventional aerofoil is best. It gives a higher maximum lift coefficient than either the flat plate or the cambered one, and the tangent (drawn with a broken line) shows that it can give a higher lift/drag ratio than either of the others. It is a lot better than the flat plate but only a little better than the cambered one. The maximum lift coefficient is $1 \cdot 3$ for the traditional aerofoil and $1 \cdot 1$ for the cambered plate. The maximum lift/drag ratio is 15 for the traditional aerofoil and 12 for the cambered plate. Different figures would have been obtained if a different traditional aerofoil profile had been used, or if the plate had been cambered more or less, but it appears that better results can be obtained at this Reynolds number, with traditional aerofoils than with cambered (or flat) plates. The same is true at higher Reynolds numbers.

At a Reynolds number of 42,000 flat and cambered plates perform a little less well than at 168,000. In each case the lift coefficient is just a little less at the lower Reynolds number, for any given drag coefficient. However, the traditional aerofoil performs so much less well that it is the worst of the three profiles instead of the best. The cambered plate gives both the maximum lift coefficient (nearly $1 \cdot 1$) and the maximum lift/drag ratio (11).

The drastic change in the performance of a conventional aerofoil occurs suddenly at a critical Reynolds number which

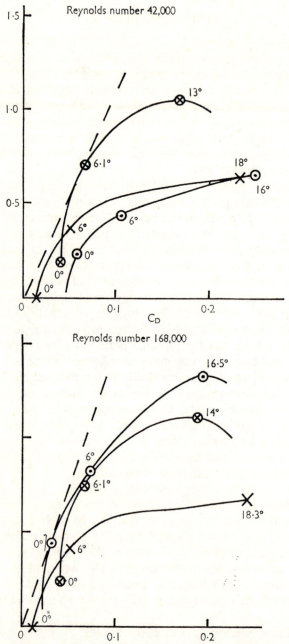

Figure 98. Polar diagrams for aerofoils of three different profiles, at Reynolds number of 42,000 and 168,000. ×, flat plate; ⊗ cambered plate; ⊙, traditional aerofoil. (Data from Schmitz, 1960)

depends both on the profile and the angle of attack, but seems generally to lie between 60,000 and 150,000.

The relative merits of profiles of different thicknesses and with different degrees of camber have been studied at the high Reynolds numbers at which aeroplane wings work (Tietjens, 1957) but apparently not at lower Reynolds numbers. A fairly strongly cambered profile gives the highest maximum lift coefficient, but less strongly cambered ones give the highest lift/drag ratio.

An aerofoil must be smooth if it is to give a high maximum lift coefficient at high Reynolds numbers, but the disadvantage of roughness becomes much smaller and eventually negligible as the Reynolds number is reduced (Schmitz, 1960).

Aerofoils do not work satisfactorily at very low Reynolds numbers. At Reynolds numbers below 100 the drag coefficients of aerofoils rise, like the drag coefficients of other bodies (Fig. 90). Though high lift coefficients are possible at Reynolds numbers around 1 the corresponding drag coefficients are so high that the lift is never more than a small fraction of the drag (Thom and Swart, 1940).

The lift on an aerofoil is due to pressure differences between the upper and lower surfaces of the whole aerofoil, but it can be represented by a force acting at a single point which is known as the centre of lift. The centre of lift may be well forward or far back depending on the profile, on the angle of attack and on whether the wing is swept back (Prandtl, 1952) but it is generally about a quarter of the way back from the front edge of a straight aerofoil.

Pennycuick (1967) needed to know the positions of the centres of lift of pigeon wings, when he calculated the lift that could be borne without breaking the wing bones (page 131). Fig. 54 (page 131) shows how he made his estimates. He took an outline of a pigeon wing and divided it into ten rectangular strips. He assumed that the lift on each strip acted half-way across it and a quarter of the way back from the front edge. This point is marked by a dot on each strip in the figure. When a pigeon is gliding all the strips have the same velocity and the lift on each strip should be proportional to its area. Pennycuick assumed this was so, and calculated that the resultant of the lifts on the individual strips would act at the point marked by the hollow star. This is the centre of lift for gliding. When a pigeon hovers the lift on a strip is no longer proportional to its area, because the distal strips move fastest and the proximal ones hardly move at all. Pennycuick allowed for this and calculated that the centre of lift for hovering would be at the point marked by the black star.

Insect wings

Insect wings work at low Reynolds numbers. Locusts fly at 350 cm/s or a little more (page 260). Their hind wings taper, but the chord is about 2 cm near the base of the wing. The kinematic viscosity of air is $0 \cdot 14$ (Table 4, page 210) and the base of the wing travels at the same speed as the body so the Reynolds number, at the base of the wing, is about $350 \times 2/0 \cdot 14 = 5,000$. Locusts are larger than most insects so their wings work at higher Reynolds numbers than most. Insect wings work at Reynolds numbers far too low for conventional aerofoil profiles to be effective. They do not, indeed, have such profiles but they are thin and sometimes cambered. Their surfaces are not smooth but they have veins rising from them as ridges. These ridges are probably too low to have an appreciable effect on either lift or drag, at such low Reynolds numbers.

Fig. 96 includes polar diagrams for a locust wing and a fruit fly (*Drosophila*) wing, at appropriate Reynolds numbers. *Drosophila* is, of course, a very small insect. The measurements were made in wind tunnels, with actual wings. The locust wing has the properties one would expect but the *Drosophila* wing seems impossible to stall. The lift coefficient remains more or less constant at angles of attack between 20° and 50°. This is not simply due to the low Reynolds number because model wings cut from recording tape stall at about 25°, at the same Reynolds number. It must be something to do with the structure of the wing: it may be due to its various hairs and bristles (Vogel, 1967).

Lift and drag on butterfly wings have also been measured by Nachtigall (1967).

Dolphin flukes

The tail flukes of dolphins and other whales are hydrofoils. The tail is moved in such a way that the lift on the fluke has a forward component which drives the animal through the water (Purves, 1963). We have seen that the dolphin *Tursiops* can swim at speeds up to 830 cm/s. Its tail fluke tapers towards the tips but has an average chord length of around 15 cm. Since the tail moves up and down as the animal swims the fluke takes a sinuous path through the water and must travel rather faster than the body, but we will ignore this in calculating the Reynolds number. The fluke has to work at Reynolds numbers up to at least $15 \times 830/0 \cdot 01 = 1 \cdot 2 \times 10^6$. This is well up in the range of Reynolds numbers in which conventional streamlined profiles are better than cambered plates. The flukes of dolphins have rather thick, streamlined profiles, very like the profile illustrated in Fig. 97c (Lang, 1966).

Bird wing profiles

Birds are larger than insects and they fly faster but their wings still work at quite low Reynolds numbers. For instance, the chord of a pigeon wing is about 12 cm, except near the wing tip. The wings of a pigeon gliding at 15 m/s are working at a Reynolds number of $12 \times 1,500/0 \cdot 14 = 130,000$. Pennycuick (1967) has calculated that the wings of hovering pigeons work at Reynolds numbers around 40,000. Pigeon wings apparently work over the range of Reynolds numbers in which the properties of a conventional aerofoil change. Wings with a conventional aerofoil profile might be best for a pigeon when it glides or flies fast, but thin cambered plates would be better when the bird hovers.

Fig. 99 shows profiles of various parts of a pigeon wing. They were obtained by Nachtigall and Wieser (1966), who used a method designed to record the profile without disturbing the feathers. The proximal end of the wing has a rather streamlined profile with a thick anterior edge which contains the bones and muscles. The covert feathers pad the wing out to a smooth profile. Distally the wing is a thin, cambered plate.

These are profiles of a wing which was stationary, in still air. In flight, air pressure bends the wing and alters the profile. Nachtigall and Wieser investigated this by fixing pigeon wings in wind tunnels. They put lines of small marks on each wing, and photographed them at various wind speeds and

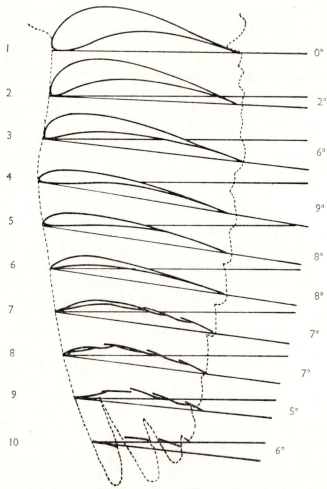

Figure 99. A pigeon wing, showing some sections. (From Nachtigall and Wieser, 1966)

various angles of attack. They used three cameras pointing at a wing from different directions and used the stereoscopic effect to calculate the position of each mark, and so the shape of the profile. They found, as one might expect, that the wind bent up the rear edges of the wings reducing the angle of attack and the camber.

A wing of which the camber decreases as the speed increases seems admirable. At low speeds a high lift coefficient is needed, if the lift on the wings is to support the weight of the body. Large camber helps to make this possible. At high speeds the lift can probably be obtained with less drag if the camber is not too great.

Brown (1953) fixed the primary feathers of a pigeon in a wooden holder which held them in the positions they would have in a spread wing. He measured the lift on them in a wind-tunnel at various wind speeds. So long as the angle of attack was not too great, lift was more or less independent of speed. This must be due to the feathers being bent more at the higher speeds.

Slotted wings

Ordinary aerofoils do not normally develop lift coefficients more than about 1·5. This is because they stall. Higher lift coefficients can be obtained from wings with devices which prevent stalling until an unusually high angle of attack is reached. The best known of these devices are slots.

Fig. 97e (page 232) shows the profile of a wing with a leading-edge slot. Wings like this give lift coefficients of 2 and more (Tietjens, 1957). Fig. 97f shows the profile of a multi-slotted wing. Wings like this give lift coefficients up to about 3·9 (Landolt and Börnstein, 1955).

The alula

The alula or bastard wing is a group of feathers which rests on the anterior dorsal surface of the wing. It can move independently because it is supported by the thumb, while the primary feathers are borne on the next two fingers. The alula can be raised so that there is a gap between it and the rest of the wing, and a section through the wing at that level looks rather like Fig. 97e. It rises at certain stages of the wing beat in slow flight (Fig. 105, page 252) and probably acts as a leading edge slot. It is some distance from the tip of the wing but the primary feathers tend to separate at the wing tip, forming in effect a multi-slotted wing.

Brown (1963b) obtained evidence that the alula works as an automatic slot. He fixed the wing of a freshly killed pigeon in a wind tunnel. It was necessary to use a freshly killed one because the experiment would not work if the muscles had stiffened. At low angles of attack the alula rested on the main part of the wing leaving no slot. As the angle of attack was increased, a point came when the alula rose. There is no mystery about this. Aerofoils give lift because the pressure is less above them than below them. At high angles of attack the pressure is likely to be particularly low just above the alula (Tietjens, 1957) so that the alula is sucked upwards.

If the alula is sewn down so that it cannot rise, the feathers on the dorsal side of the wing start to flutter at an angle of attack just above the one at which the alula would have risen. The fluttering indicates that the wing has stalled. If the stitches are then cut without moving the wing the alula rises and the feathers lie smoothly again, showing that the wing is no longer stalled. This seems a convincing demonstration that the alula does indeed delay stalling, as one would expect a leading-edge slot to do. However, Nachtigall and Wieser (1966) were unable to demonstrate automatic rising of the alula in their experiments.

Gliding

Fig. 100a represents a glider travelling at a constant airspeed u at a gliding angle θ. The airspeed is the velocity relative to the

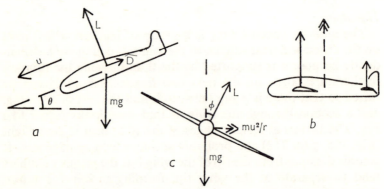

Figure 100. Diagrams of gliders, (*a*) shows the forces which act in a straight glide, (*b*) illustrates the account of the function of elevator flaps, and (*c*) shows the forces which act in a horizontal turn

air and the gliding angle is the angle relative to the horizontal at which the glider travels through the air. If there is no wind, they will be the same as the velocity and angle relative to the ground. The glider is in equilibrium under the action of the lift L (due mainly to the wings) the drag D and the force due to gravity, mg. Therefore

$$L = mg \cos \theta \qquad (33)$$

and
$$D = mg \sin \theta \qquad (34)$$

Equation (31) states that

$$L = \tfrac{1}{2}\rho u^2 A C_{\mathrm{L}}$$

where ρ is the density of the air, A is the area of the wings and C_{L} is the lift coefficient.

From equation (32)

$$D = \tfrac{1}{2}\rho u^2 A(C_{\mathrm{DO}} + kC_{\mathrm{L}}^2/\pi R)$$

where C_{DO} is a profile drag coefficient which takes account of the drag on the fuselage as well as the drag on the wings, and R is the aspect ratio.

These equations can be used to obtain an equation for the speed, $u \sin \theta$, at which the glider sinks relative to the air. This is called the sinking speed. Provided the gliding angle is small, as it would normally be,

$$u \sin \theta = (\rho A C_{\mathrm{DO}}/2mg)u^3 + (2kmg/\pi R\rho A)/u \qquad (35)$$

(Welch, Welch and Irving, 1955). Notice that one of the terms on the right-hand side of the equation is large when u is large, and the other is large when u is small. There is an intermediate value of u at which $u \sin \theta$ is least: there is an optimum airspeed which gives a minimum sinking speed. Fig. 101 shows graphs of sinking speed against airspeed for a glider and a bird. The one for the glider was obtained by test flying. The other, for the fulmar, was obtained by a method which will be described in the next section of this chapter. Both have the form predicted by equation (35) (note that the scale of sinking speed reads downwards).

An increase in aspect ratio always reduces the sinking speed, but is very much more effective at low speeds than at high ones

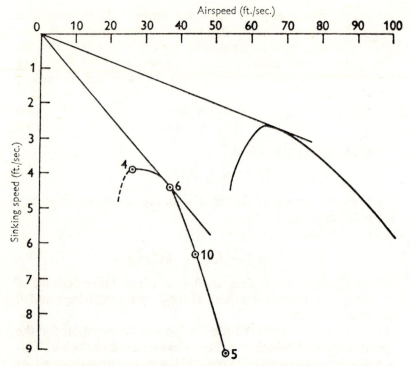

Figure 101. Graphs of sinking speed against airspeed for a typical glider (right) and for gliding fulmars (left). The numbers beside the latter curve show the number of observations on which each point is based. (From Pennycuick, 1960)

because the term it affects in equation (35) is the one that is large when u is small. An increase in wing area may either increase the sinking speed or reduce it. It can be shown by calculus (by differentiating equation (35) with respect to A) that the sinking speed is least for a given airspeed when

$$A = 2mg\sqrt{[k/\pi RC_{\text{DO}}]/\rho u^2} \qquad (36)$$

so the ideal wing area is large at low speeds and small at high speeds. The ideal wing loading (m/A) is low at low speeds and high at high speeds.

Equation (35) shows how airspeed depends on gliding angle, for any particular glider. The gliding angle can be adjusted, by moving the elevator flaps. The lift on the glider is the resultant of the lift on the wings and the lift on the elevators (Fig. 100b). The elevator flaps are strips along the rear edges of the elevators, which can be hinged down (as in Fig. 100b) to increase the lift on the elevators or up to reduce it. When the lift on the elevators is increased, the resultant of the lift on the wings and the lift on the elevators is shifted back. When the lift on the elevators is reduced, the resultant lift is shifted forward. These changes alter the gliding angle because the glider can only be in equilibrium, when the point where the lift and drag intersect is vertically above the centre of gravity.

When an aircraft turns, it must bank: it must tilt towards the inside of the turn so that the lift has a horizontal component to provide centripetal force. If the mass of the aircraft is m a centripetal force mu^2/r is needed for a turn of radius r at a speed u (see page 33). This is illustrated in Fig. 100c, which represents an aircraft making a turn while flying level. The vertical component of the lift is $L \cos \phi$ and it must balance the weight so

$$mg = L \cos \phi$$

The horizontal component is $L \sin \phi$ and it must provide the centripetal force so

$$mu^2/r = L \sin \phi$$

From these two equations we find

$$m^2g^2 + m^2u^4/r^2 = L^2(\sin^2 \phi + \cos^2 \phi) = L^2$$
$$L = m\sqrt{[g^2 + u^4/r^2]} \qquad (37)$$

The greater the speed and the smaller the radius of the turn, the more lift is needed. The equations also give us the angle of bank

$$\tan \phi = u^2/rg \qquad (38)$$

An aircraft can be made to turn, by making it bank. This is done by means of ailerons, which are flaps on the rear edges of the wings. To turn right, the left aileron is turned down and the right one up. This makes the left wing give more lift than the right one and makes the aircraft bank in the required direction.

Air-brakes are another control provided in gliders. They are plates which can be extended above and below the wings, at right angles to the path of the glider. When they are extended considerable pressure drag acts on them, slowing the glider down. Air-brakes make it possible to fly slower than the airspeed predicted by equation (35), at any particular gliding angle.

Gliding of fulmars

The fulmar (*Fulmarus*) spends a lot of time gliding in front of the cliffs where it breeds, and can be watched at close quarters from the cliff top. Pennycuick (1960) exploited this in a study of its gliding flight. He measured the airspeeds and gliding angles of a lot of gliding fulmars.

He set up a camera on the top of the cliff, on a tripod with a pan-tilt head which allowed the camera to be pointed in any direction. The head was connected to a pen recorder so that one pen drew a continuous record of the compass direction in which the camera was pointing and another drew a record of the angle at which it was tilted. A rangefinder was mounted on the camera and was connected to a third pen which recorded the range to which it was adjusted. The pen recorder was started as a bird approached and the camera was kept trained on the bird with the rangefinder adjusted to its range. The pen recorder made a record of the direction and distance of the bird, from which its path and speed relative to the ground could be measured. The speed and direction of the wind were measured by a wind meter, held over the cliff on the end of a pole. The speed and direction of the bird relative to the air were obtained from these data.

A device fitted to the rangefinder fired the camera automatically when the bird reached a distance of 35 ft. The photograph was used to estimate the wing area, which the bird can vary by bending and extending the wing. Fig. 102 shows outlines of the wing of a dead fulmar, in various positions. The wing areas of the gliding fulmars were estimated by comparing the photographs with these outlines. Adult fulmars are reasonably uniform in size.

Fulmars can retract their feet so that they are completely hidden by feathers, and they do this when they are gliding steadily in smooth conditions. They sometimes lower their feet and use them as air-brakes so that they can, for instance, make a steep descent reasonably slowly.

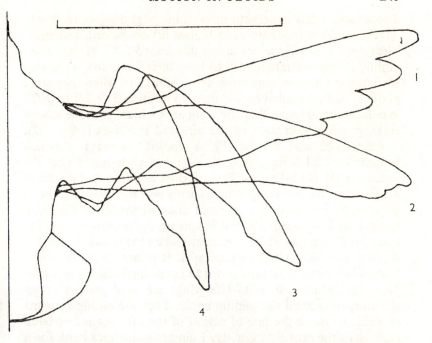

Figure 102. Outlines of a dead fulmar showing how the areas of the wing
and tail can be varied. Scale: 1 foot. (From Pennycuick, 1960)

Fig. 101 has already been introduced. The curve for fulmars
is based entirely on records of glides with the feet completely
hidden. The tangent shows that the minimum gliding angle is a
little less than 7°. The minimum sinking speed seems to be about
4 ft/s. In both respects the bird is inferior to the glider whose
performance is shown in the same figure. Pennycuick points out
that a glider is designed solely for gliding while "a gliding fulmar
is not a glider but a powered aircraft with its motor idling, and
in this context its performance is by no means contemptible".

Pennycuick calculated lift coefficients from his records of
gliding fulmars. The apparatus did not provide a reliable record
of acceleration so he assumed that all the birds were gliding
with uniform velocity and used equation (33) (page 241) to
calculate the lift. He found a wide range of lift coefficients up to
a maximum of 1·8, which is just enough to support a fulmar
gliding at 23 ft/s with its wings extended in position 1 (Fig. 102).

9—AM

There were a few observations of fulmars gliding more slowly which would indicate an even higher lift coefficient, but these birds were probably accelerating downwards. When they were gliding slowly fulmars tended to keep their wings fully extended so that their area was as large as possible, but when they were gliding fast they tended to fold them to position 2 or 3 (Fig. 102). We have already seen that the wing area which gives the lowest sinking speed, decreases as the airspeed increases (page 242).

Pennycuick and Webbe (1959) studied the ways in which fulmars control their gliding. We have already noted that the feet are used as air-brakes. The tail does not seem to be used as an elevator, for moulting fulmars with no tail feathers seem to manoeuvre just as well as normal ones, at moderate and high speeds. At low speeds the tail is spread, apparently to provide extra lift (frontispiece). This happens when the fulmar lands, for it must slow down before landing if it is not to injure itself. Some birds bend their tails down to act as air-brakes when they land, but fulmars do not. The wings are used instead of an elevator, to control the gliding angle. They are swung forward or back, to move the line of action of the lift forward or back relative to the centre of gravity. Fulmars sometimes bank for a turn by rotating their wings about their long axes, so that they have different angles of attack (frontispiece): the whole wing is used as an aileron. Sometimes they only twist the tips of the wings.

Thermal soaring

Thermals are lumps of air that rise where the earth is heated irregularly by the sun. Over a part of the ground that becomes hotter than its surroundings, the air gets warmer and less dense than the surrounding air and eventually rises as a "thermal". A thermal often lasts for about 5 minutes and is followed by an interval of 15 minutes or so. There are often thermals over dry, dark areas of ground and over slopes that face the sun. Glider pilots study both the ground and the clouds when they are looking for thermals (Welch, Welch and Irving, 1955).

A glider in a thermal will rise, if its sinking speed is less than the upward velocity of the air. It may gain a lot of height by circling for some time in the thermal and can then glide along a straight path for some distance, losing height, before it is necessary to find another thermal and circle again.

A glider in a small thermal must circle with a small radius. The smaller the radius, the more lift is needed (equation 37). Extra lift can be obtained by flying faster, but this increases the lift needed for circling still more. Extra lift can be obtained, within limits, by increasing the angle of attack and so the lift coefficient, but this increases the induced drag and so the sinking speed. Welch, Welch and Irving (1955) considered the glider whose performance is shown in Fig. 101 (page 242) and showed that its minimum sinking speed would increase rapidly as the radius fell towards 100 ft. At a radius of 100 ft the sinking speed would be 20 ft/s and even the strongest thermals would be too slow to make the glider rise. The glider can only soar in thermals which are several hundred feet across.

Soaring buzzards

Rooks, gulls and various other birds soar in thermals, circling like gliders (Welch, Welch and Irving, 1955). Raspet (1960) used a glider to study the soaring and gliding of *Coragyps*, the black buzzard. His glider was highly manoeuvrable and he had accurate knowledge of its performance. He followed buzzards at a distance of 5–10 m, so that the airspeed of the glider (which could be read off the airspeed indicator) was the same as the airspeed of the buzzard. He knew the sinking speed of the glider at any airspeed and determined the sinking speed of the buzzard, by observing how fast it rose or sank relative to the glider. The buzzard's performance was very like the performance of the glider shown in Fig. 101. The minimum gliding angle was 2° 30′, while that of the glider was 2° 18′. The minimum sinking speed was 0·6 m/s while that of the glider was 0·8 m/s, but the bird had its minimum sinking speed at a lower airspeed (14 m/s) than the glider (19 m/s). Since the extra lift needed for circling increases with speed (equation 37) the buzzard could presumably soar in smaller thermals than the glider.

When buzzards soar or glide slowly they splay apart the primary feathers of their wings (Fig. 103). This makes the ends of the wings into multi-slotted aerofoils (see page 239) and

Figure 103. Sketches of a black buzzard (*Coragyps*) showing the positions of wings in soaring (left) and in fast gliding. (Redrawn from Newman, 1958)

makes possible the high lift coefficients which are needed at low airspeeds. It has often been stated that separation of the primaries reduces induced drag by making the wing behave as though it had a bigger aspect ratio, but this idea seems to be wrong (Newman, 1958).

When buzzards glide fast they narrow or close the gaps between the primaries, and reduce the wing area by flexing the wings. The advantage of reducing the wing area at high speeds has been explained (page 242).

Albatrosses

Albatrosses (*Diomedea*) can be watched by the hour gliding over the sea, rising to about 12 m and swooping down close to the surface, hardly ever flapping their wings (Jameson, 1958). They are able to stay airborne in this way, because the speed of the wind is less, close to the surface of the sea, than it is higher up.

A bird that has got up speed can always glide upwards for a while, until it is slowed down by gravity and by drag so much that its wings cannot provide enough lift. An albatross faces into the wind as it rises, and its speed relative to the sea decreases as it rises, but its airspeed may remain constant or even increase since the speed of the wind it is facing increases as it rises.

Walkden (1925) discussed the mechanics of albatross soaring. He estimated that an albatross gliding with an airspeed u m/s could only glide upwards without losing airspeed if the speed of the wind increased by at least $13/u$ m/s for every metre increase of height. The faster the bird glides, the smaller is the gradient of wind speed needed for soaring. Albatrosses glide at airspeeds up to about 28 m/s so the minimum gradient must be about $0\cdot5$/s. Gradients of this order are only found near the surface of the sea and this is presumably why albatrosses never soar very high.

Once it has reached the top of its climb an albatross may glide down again in any direction. It may continue against the wind and if it is gliding fast enough may actually make headway against the wind, although it is not flapping its wings. Alternatively, it may turn and glide downwind, in which case it will travel some distance downwind during each cycle of rising and falling.

This kind of soaring is very different from the thermal soaring of buzzards. An albatross can only soar in weak wind gradients by travelling at high airspeeds. A buzzard can only soar in a small, weak thermal if it can glide at low airspeeds. The wings of an albatross and a vulture are compared in Table 5. Vultures soar in the same way as buzzards, and a vulture has been used in preference to the much smaller black buzzard discussed in the preceding section of this chapter, so that birds of similar size can be compared.

TABLE 5

THE WING LOADING OF AN ALBATROSS (*Diomedea exulans*) AND A VULTURE (*Gyps fulvus*) OF SIMILAR SIZE

(Data from Greenewalt, 1962)

	Weight (gm)	Wing span (cm)	Wing area (cm²)	Wing loading (gm/cm²)
Albatross	8,502	341	6,206	1·37
Vulture	7,269	256	10,540	0·69

The albatross has rather longer wings than the vulture but they are so much narrower that the wing loading is twice as high. A fairly high wing loading is best for the albatross, when it glides fast (equation 36, page 242). A low wing loading is essential for the vulture if its wings are to provide enough lift at low speeds. High aspect ratios would be advantageous for both birds, but long wings have large bending moments at their bases and a very long wing would need a very heavy skeleton. The vulture has to tolerate a relatively low aspect ratio to get a big enough wing area without having too long a wing. The aspect ratio of the albatross is about 18, but that of the vulture is only about 6. The primary feathers of vultures and buzzards separate to form multi-slotted aerofoils to maintain lift at low speeds, but the primaries of albatrosses do not.

Helicopters

The blades of a helicopter rotor are aerofoils which provide lift like ordinary aeroplane wings, but they can provide it even when the aircraft is stationary because they move through the air as the rotor rotates. When air exerts an upward lift on an aerofoil the aerofoil exerts an equal downward force on the air, and the air is accelerated downwards. By knowing how much work must be done accelerating air downwards, one can calculate the minimum power needed to keep a helicopter stationary in the air (Prandtl, 1952). If the mass of the helicopter is m and if the area of the circle swept out by the rotating blades is A, the minimum power Q needed in air of density ρ is given by the equation

$$Q = \sqrt{[m^3 g^3 / 2A\rho]} \tag{39}$$

This is the power used to accelerate the air downwards. It must be divided by the hydraulic efficiency of the rotor (see page 223) to obtain the power needed to drive the rotor.

Humming-birds

Humming-birds feed on nectar, which they take from flowers without landing. They hover in front of the flowers, beating their wings forward and back as shown in Fig. 104. Notice that the wing is turned upside-down as it is swung back. It works like a blade in a helicopter rotor, except that it keeps changing direction. The turning over keeps the lift acting upwards all the time. Each wing swings forward and back through about 120° (Hertel, 1966).

We can estimate the power needed for hovering, by means of equation (39). We will consider a fairly small humming bird, weighing 3 g. Its wings would be about 4 cm long (Greenewalt, 1962) and since each swings forward and back through 120° the area A would be 2/3 of the area of a circle of radius 4 cm, or $16\pi \times 2/3 = 34$ cm². The density of air is about $1\cdot3 \times 10^{-3}$ g/cm³. Putting these values in equation (3) we find

$$Q = \sqrt{[(3 \times 981)^3/2 \times 34 \times 1 \cdot 3 \times 10^{-3}]}$$
$$= 5 \cdot 4 \times 10^5 \text{ erg/s}$$

Pearson (1950) and Lasiewski (1963) measured the rates at which hovering humming-birds used oxygen. The bird was put in a large airtight chamber. A little cup of sugar solution was hung in the chamber in such a way that the bird had to hover to feed from it. Air was pumped slowly through the chamber and the incoming and outgoing air were analysed. A 3 g *Calypte costae* hovered continuously for 35 min in one of Lasiewski's

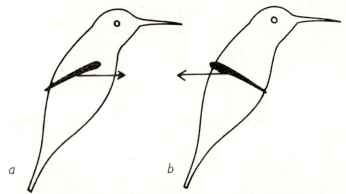

Figure 104. Diagrams showing how a hovering humming-bird moves its wings. The black areas are sections of a wing

experiments, using oxygen at an average rate of 127 ml/h. It only used about 18 ml oxygen/h when it was resting in the dark, so the oxygen needed for the activity of hovering must have been about 110 ml/h. This would release about 550 cal/h of chemical energy, whether carbohydrate or fat was being oxidized (Bell, Davidson and Scarborough, 1965). This is equivalent to $6 \cdot 4 \times 10^6$ erg/s. We estimated that the power used usefully in hovering was $5 \cdot 4 \times 10^5$ erg/s so the efficiency of conversion of chemical energy to useful work was apparently about 8%. The efficiency of conversion of chemical energy to mechanical work by a man pedalling a bicycle is about 20% at the optimum rate of pedalling (page 24). If 20% of the chemical energy used by the humming-bird muscles was converted to mechanical work the total mechanical work done by the muscles would have been $0 \cdot 2 \times 6 \cdot 4 \times 10^6 = 1 \cdot 3 \times 10^6$ erg/s. Since $5 \cdot 4 \times 10^5$ erg/s was useful work, the hydraulic efficiency of the hovering mechanism seems to be about 40%.

The wing muscles of a 3 g humming-bird would weigh about $0 \cdot 75$ g (Greenewalt, 1962) so the mechanical power output per unit weight of muscle would be $1 \cdot 7 \times 10^6$ erg/s g or 170 W/kg. This is about 4 times the maximum power output of human muscle (page 24).

Flapping flight of birds

Though we have discussed gliding and hovering we have still not dealt with ordinary forward flapping flight. Brown (1948,

Figure 105. Outlines traced from photographs taken at intervals of $0 \cdot 01$ s, showing a pigeon flying slowly. The sequence is continued in Fig. 106. (From Brown, 1963b)

1953, 1963b) studied it by cinematography, especially of pigeons. Pigeons may make as many as 9 wing beats per second (Pennycuick and Parker, 1966) so he took his films at about 90 frames/s.

Figs. 105 and 106 show a pigeon flying very slowly, just after take-off. Slow flight seems much more strenuous than normal

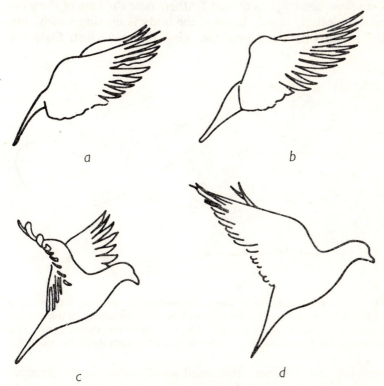

Figure 106. These outlines complete the sequence started in Fig. 105. (From Brown, 1963b)

flight and can only be maintained for a few seconds. The body is tilted at a large angle to the horizontal. In the downstroke the alula is open. The wings are swung downwards and then forwards: they extend laterally from the body in Fig. 105c but are nearly parallel to each other in front of the body in Fig. 105d. Fig. 106a shows the beginning of the upstroke with the wing hiding the head. The wing moves upwards and back led by the bent wrist until there is a sudden flick which moves the wing tip very rapidly through the position shown in Fig. 106c and extends the wing above the back. Brown (1948) published a series of diagrams showing successive positions of the wing bones in this complicated movement.

In slow flight the secondary feathers near the base of the wing move relatively slowly because the body is moving slowly, but the primary feathers near the wing tip move fast. Only the

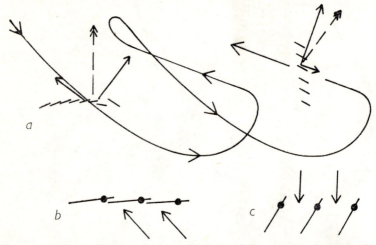

Figure 107. (*a*) A diagram showing the path of the primary feathers of a pigeon flying slowly and the forces which act on the primary feathers in the downstroke and in the upstroke. (*b*), (*c*) Diagrams showing why the primaries keep together in the downstroke and separate in the upstroke

primaries can produce substantial aerodynamic forces. Removing the secondary feathers seems to make no difference to a pigeon's ability to fly slowly, but removing part of the distal primaries makes the bird unable to fly. We need only consider the primaries in explaining slow flight.

The primaries move rapidly down and forward and then up and back, relative to the bird, while the bird moves slowly through the air. The path of a point on a primary feather relative to the air is the sum of these two movements, more or less as shown in Fig. 107a. This figure also shows sections of the distal part of the wing. During the downstroke it has a large angle of attack but is prevented from stalling by slotting (see page 239). The open alula acts as a leading-edge slot and the primaries separate a little at the wing tip, making it multiply slotted. The resultant of lift and drag probably acts roughly vertically upwards, as is indicated by the way the primaries are bent in Fig. 105*c*.

In the upstroke the pressure of air on the dorsal surfaces of the primaries twists them, so that they are not merely splayed apart at their ends as in the downstroke, but are separated along most of their length. This is because the vane of each feather is wider on the posterior side of the shaft than on the anterior side, and because the feathers overlap as shown in Fig. 107*b*. Air pressure on the wider part of the vane tends to keep the feathers together during the downstroke (except at their tips, where they do not overlap), but it twists them apart in the upstroke (Fig. 107*c*). It has been suggested that the downstroke

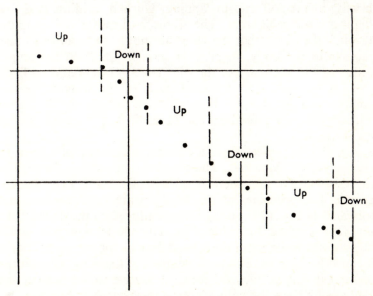

Figure 108. Successive positions at equal intervals of the shoulder of a pigeon, flying upwards and to the left. Note that the speed is greater in the upstroke than in the downstroke. (From Brown, 1953)

is the power stroke and the upstroke is merely a recovery stroke with the feathers "feathered" like oars, but the primaries are bent up and forward in Fig. 106c and this shows that an upward and forward force is acting on them. Fig. 107a shows how the primaries act as individual aerofoils and produce this force.

A flying bird needs upward aerodynamic forces to support its weight and forward ones to drive it forward. Our analysis so far indicates that the resultant force on the wing acts more or less directly upwards in the downstroke, but upwards and forwards in the upstroke. This seems to be confirmed by Fig. 108, which shows the position of the shoulder of a pigeon in successive frames of a cinematograph film. The bird is flying slowly, climbing from the right to the left. The points are more widely spaced during the upstrokes than during the downstrokes, so the bird travels faster during the upstrokes in which we believe the main propulsive force to act.

The slow flight we have considered so far is only used very briefly, at take-off and in landing. Normal, medium and fast flight are quite different. The whole bird moves quite rapidly through the air so that the wings do not have to move so fast, relative to the body, to provide enough lift. Both the amplitude and the frequency of the wing beats is less in normal flight than at take-off. The proximal parts of the wing move fast enough through the air to provide useful forces. A pigeon without secondary feathers can take off, but it cannot get up much speed.

Fig. 109 was traced from a cinematograph film of a great tit (*Parus major*) flying fast. The wings are not swung forward in the downstroke, and there is no complicated flick in the upstroke. The primaries at the wing tip are bent upwards and forwards in Fig. 109c, so the force produced by the downstroke must be inclined forwards. Fig. 110 shows how this could happen: the resultant of the lift and drag on the wing is of course inclined backwards relative to the path of the wing through the air, but it may be inclined forward relative to the path of the bird's body since the path of the wing slopes.

The primaries are not bent in the upstroke so any forces that act on them are probably small. The wing tip probably rises

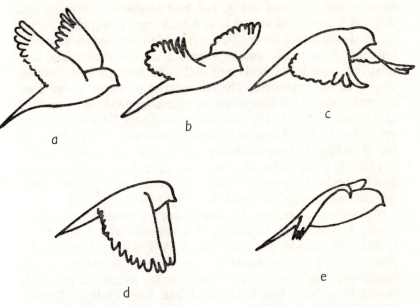

Figure 109. Outlines traced from photographs taken at intervals of $0 \cdot 01$ s, of a great tit (*Parus*) flying fast. (From Brown, 1963b)

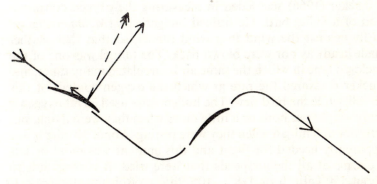

Figure 110. A diagram showing the path of the primary feathers of a bird in normal fast flight, and the forces which act on the primary feathers in the downstroke

more or less as shown in Fig. 110, and produces little or no lift. Any lift it produced would have a backward component, tending to slow the bird down. This is true of the secondaries as well, but only to a much smaller extent since the secondaries are relatively close to the body and so do not move up and down as steeply as the primaries. The secondaries probably provide an upward force all the time while the primaries produce an upward and forward force in the downstroke and very little force in the upstroke. A film of a swan flying in essentially the same way as the tit in Fig. 109 shows clearly that it accelerates in the downstroke when the primaries produce their forward force, and decelerates in the upstroke. There are modes of flight intermediate between typical fast and slow flight (Brown, 1953).

We will now estimate the power which is needed for flapping flight. A bird which is gliding at a constant speed in still air loses height, and so potential energy. A bird flying horizontally does not. We may reasonably suppose that the energy needed for flapping flight must be at least as much as the energy which would have been lost if the bird had been gliding (Raspet, 1960). The minimum sinking speed of a gliding fulmar is about 120 cm/s (page 245) and that of a black buzzard is 60 cm/s (page 247). We will use an intermediate value of 100 cm/s and reckon that a gliding bird of mass m g will generally lose 100 mg erg potential energy/s, and that a bird will need at least 100 mg erg/s for level flight.

Tucker (1966) succeeded in measuring the oxygen consumption of a flying bird. He trained budgerigars (*Melanopsittacus*) to fly against the wind in a wind tunnel, so that they neither made headway nor were blown back. The tunnel was one of the enclosed type in which the same air is circulated many times, and Tucker measured the rate at which the oxygen content of this air fell, while the bird flew. The budgerigars used 38 ml oxygen/g body weight per hour or a little more when they were flying, but only about 8 ml/g h when they were resting. About 30 ml/g h was apparently needed for flight and this amount was more or less the same at all the airspeeds that were tried. It corresponds to about 150 cal/g h or $1 \cdot 8 \times 10^6$ erg/g s of chemical energy. If the efficiency of conversion of chemical to mechanical energy in the muscles is 20% (see page 24), about $3 \cdot 6 \times 10^5$ erg of mechanical energy was released per gram body weight per second. We estimated from the observations of fulmars and black buzzards that 100 mg erg/s or 10^5 erg/g s would be needed. This suggests that the hydraulic efficiency of flight is around 30%, which is a little less than our estimate for the hovering humming-bird (page 252). The calculation is of course a very rough one and it must be admitted that neither a fulmar nor a buzzard is particularly like a budgerigar.

The power was given in terms of the weight of the whole body, not just of the muscles. The flight muscles of a budgerigar are 26% of the weight of the body so the mechanical power output estimated from the oxygen consumption is about $1 \cdot 4 \times 10^6$ erg/g muscle s or 140 W/kg muscle, which is nearly as high as the estimate for the humming-bird.

Insect flight

It might be possible to study the flight movements of insects as Brown (page 252) studied the flight movements of birds, by taking cinematograph films of insects flying freely. Weis-Fogh (1956a) devised a different method which he used to study the flight of locusts.

A locust was fixed to a slender lever, like the bob on a pendulum. The attachment was made by wax or a suction cup, in such a way as not to interfere with wing movements. It was then suspended in front of the nozzle of a wind tunnel like the one shown in Fig. 89 (page 214). When the wind was started it would flap its wings as in flight and might "fly" for several hours. If it flew faster than the wind it would pull the pendulum forwards and if it flew slower it would be blown back. The pendulum was connected electrically to the fan of the wind tunnel, so that the wind speed was increased when the pendulum swung forwards and reduced when it swung back. In this way the wind speed was automatically adjusted to be exactly equal to the speed at which the locust was "flying". The backward force due to drag was then equal to the forward force due to the flight movements, as it would be in free flight at a constant speed. Part of the drag was, of course, drag on the slender arm of the pendulum. This was small but steps were taken in some of the experiments to counteract it.

The pendulum was suspended from an aerodynamic balance, which measured the upward force produced by the flight movements. When this force was equal to the locust's weight, the conditions of the experiment approximated to free level flight. This happened when the locust was "flying" at about 350 cm/s, but the upward force was less at lower speeds and more at higher ones. Estimates of the airspeeds of wild locusts flying freely are a little more than 350 cm/s.

The flight movements were recorded by photography. A stroboscopic method was used in preference to high-speed cinematography, which is expensive and needs strong lights (the locusts flew most steadily in subdued light). The intervals between the stroboscope flashes were adjusted to be slightly more than the period of the wing beat, so that the wings looked as if they were moving in the normal way but much more slowly. Each flash was used to take a photograph; each photograph was of a different wing beat but each showed a slightly later stage of the cycle of wing movements than its predecessor. The result was, in effect, a slow motion film of the very fast cycle of wing movements.

Before we look at Weis-Fogh's results let us examine the flight movements of *Phormia*. This is a fly which was only two wings, while locusts have four. It was studied by Nachtigall (1966) whose method was essentially the same as Weis-Fogh's method for locusts, though he used high-speed cinematography.

Fig. 111 shows how *Phormia* flies. The two sketches show it at the top of the upstroke and at the bottom of the downstroke. The wings swing forward relative to the body in the downstroke and back in the upstroke. As the body moves forward relative to the air all the time, a point on a wing moves relative to the air along a sinuous path, as shown in the figure.

Fig. 111 shows sections of the wing at four stages in the cycle of movements, and the forces that probably act on it. For most of the downstroke the anterior edge of the wing is a little lower than the posterior edge but in the upstroke the wing surface is nearly vertical and the anterior edge is uppermost. The mechanism of these movements has already been described (page 52). At stage *a* the wing has a relatively small angle of attack and presumably gives reasonable lift and little drag. The

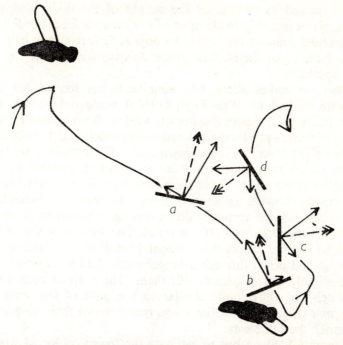

Figure 111. A diagram illustrating the wing movements of a fly and the forces which probably act on the wing at various stages in its cycle of movements. The sinuous line is the path of the wing tip relative to the air.
(Based on illustrations by Nachtigall, 1966)

resultant force is probably roughly vertical. At *b* the angle of attack is nearly 90° and the drag must be high, but if the wing is as hard to stall as the wing of *Drosophila* (Fig. 96, page 231) there may still be substantial lift. In any case the resultant force must again be roughly vertical. The high drag will help to slow down the wing, which must be halted at the bottom of the downstroke. At *c* the air strikes the dorsal surface of the wing instead of the ventral surface and the resultant force probably acts roughly horizontally, forwards. At *d* the air strikes the ventral surface again and the resultant force probably acts backwards and downwards. It can serve no useful function except that of decelerating the wing at the top of the stroke.

Thus, the downstroke seems to provide most of the upward

force needed to counteract the weight of the body, and the upstroke seems to provide most of the forward force needed to counteract drag on the body. Aerodynamic forces may help to halt the wing at the bottom of the downstroke and the top of the upstroke.

Phormia makes about 120 wing beats per second but the locusts studied by Weis-Fogh (1956a) made only about 17. Fig. 112 is a composite illustration built up from several typical films of locusts producing upward forces which exactly balanced their weight. The thin lines drawn across the forewings are thin hairs which were glued to them to make it possible to see rotation of the wings about their long axes. Below the pictures are graphs showing the angle of the wings, seen from behind. If the wings pointed vertically down this angle would be given as zero, if vertically up as 180° or π rad. The line across the graph at $1\cdot6$ ($\pi/2$) rad marks the horizontal position. The hind wings flap up and down through a bigger angle than the fore-wings, and slightly out of phase with them. The path of each wing through the air is quite similar to the path of the wing of *Phormia* (Fig. 111) and the wings rotate about their long axes in much the same way.

Jensen (1956) set out to estimate the forces on locust wings at each stage in the cycle of movements revealed by Weis-Fogh's work. He had to measure the lift and drag on the wings at various angles of attack. The tips of the wings move faster than their bases and account had to be taken of this in the analysis. Jensen simplified the analysis by measuring the lift and drag on wings in an air current whose velocity was greater at the tip of the wing than at the base. We have already seen that fluids travel more slowly near the wall of a pipe than near the centre (page 211). Jensen inserted a locust wing through a hole in the wall of a pipe through which a current of air was passing. By shaping the wall of the pipe suitably he was able to obtain a graded flow over the wing which corresponded reasonably well with conditions in flight. He measured lift and drag on each wing in a succession of positions corresponding as closely as possible to the positions occurring in flight. He used the results to estimate the vertical and horizontal forces exerted by the wings at each stage in the cycle of wing movements.

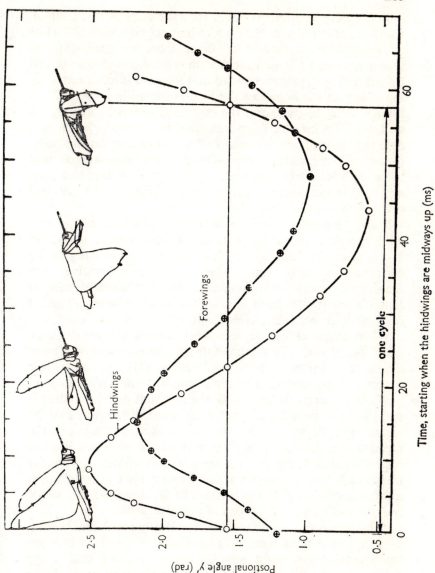

Figure 112. Graphs of the angles of the wings (in radians, relative to vertical downwards) against time, from a film of a locust "flying" in a wind tunnel. Outlines of the locust in four positions are shown above the curves. (From Weis-Fogh, 1956a)

It would be possible to doubt the soundness of this method. Can one really estimate the forces which act on a wing in the fast and complex movements of flight from measurements on stationary wings? Jensen checked the soundness of the method by estimating the mean upward and forward forces exerted by the wings on the bodies of the locusts shown in four of Weis-Fogh's films. In each case the estimated upward force agreed well with the upward force measured by Weis-Fogh. In each case the forward force agreed well with the drag on the body, measured at the same airspeed. This drag was measured on dead locusts whose wings had been removed, whose legs had been arranged in the positions they occupy in flight, and which had been dried to make them stiff.

Thus, the method seems to be reliable. The force acting on a wing at a particular stage in its movement does not always act in the same direction as the force on a *Phormia* wing at the corresponding stage, but we need not concern ourselves with the differences. The most interesting use that has been made of Jensen's results has been in calculations of the energy needed for flight (Jensen, 1956; Weis-Fogh, 1961c).

When an insect flies, it does work against the aerodynamic forces on its wings. In the downstroke, the resultant of lift and drag acts upwards on the wings (see Fig. 111) and resists their downward movement. The flight muscles must obviously do work. In the upstroke in locusts, the resultant of lift and drag is still inclined upwards, and helps the upward movement of the wings. The flight muscles are not contracting against the aerodynamic forces, but are being stretched by them. We can regard them as doing negative work. It was calculated from the results of Jensen's experiments that in every cycle of wing movements a locust's flight muscles have to do about $+8,000$ erg of positive work and $-2,500$ erg of negative work.

A flying insect also does work accelerating and decelerating its wings as it flaps them up and down. When the wings are being accelerated, flight muscles are contracting and doing positive work. When the wings are being decelerated, flight muscles are being forcibly extended and are doing negative work. The amounts of work which must be done can be calculated from the maximum angular velocity of each wing in each stroke, and the moments of inertia of the wings. We have already seen how the moment of inertia of an insect wing can be measured (page 34). It seems that a locust must do about $+6,000$ erg of positive work accelerating its wings in each cycle of wing movements, and about $-6,000$ erg of negative work decelerating them.

We will call the work done against aerodynamic forces aerodynamic work, and the work done accelerating and decelerating the wings inertial work. The total work needed for each cycle of wing movements cannot be obtained simply by adding the aerodynamic and inertial work together, as can be seen by considering stage b in the wing movements of *Phormia* (Fig. 111). At this stage the wings are moving down against the aerodynamic forces, so positive aerodynamic work is being done, but they are being decelerated, so negative inertial work is being done. The kinetic energy taken from the wings can be used to do at least some of the aerodynamic work, so the positive aerodynamic work and the negative inertial work cancel each other out, at least in part. A calculation which allows for this sort of thing indicates that a locust must do a total of $+11,000$ erg of positive work and $-6,000$ erg of negative work in each cycle of wing movements.

Elastic materials can store energy (page 69). When an elastic structure is stretched work is done on it, but it can do work in the elastic recoil. In the terms of the last few paragraphs, the structure does negative work when it is stretched and positive work during the recoil. We have already seen that most of the negative work which is done at the end of the upstroke, decelerating the wings, is stored in resilin and other elastic materials (page 79). It is released again as positive work which accelerates the wings at the beginning of the downstroke. Since the inertial work makes up so large a proportion of the total work, a substantial amount of energy can be saved in this way.

This energy would be lost if the wings were simply decelerated by the muscles, because muscles cannot store work which is done on them (except for a little in the series elastic component, see page 107). Far from storing energy, a muscle actually uses up some energy when it resists forcible extension.

Heterocercal tails

Fig. 113*a* shows a typical selachian. Its tail is heterocercal, with a large lobe ventral to the vertebral column and a small lobe dorsal to it. It moves from side to side through the water as the fish swims. Fig. 113*b* is a transverse section showing forces which act on it as it moves to the left. The big ventral lobe is bent so that it has an angle of attack α, and lift acts on it as well as drag. This lift is the upward force *C* shown in Fig. 113*a*.

Typical selachians are a good deal denser than the water they live in. They need upward forces to balance their excess weight when they swim horizontally. These forces are provided by the tail and by the pectoral fins which are held at an angle of attack so as to produce lift. This is the upward force *B* shown in Fig. 113*a*.

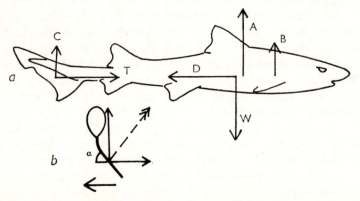

Figure 113. Diagrams showing the forces which act on a swimming selachian. (*b*) is a transverse section of the tail as it moves towards the left. (Partly after Alexander, 1967b)

The other vertical forces shown in the figure are the weight of the fish, *W*, which acts at the centre of gravity and the upthrust *A*, due to the weight of water displaced, which acts at the centre of buoyancy. The centre of gravity is slightly posterior to the centre of buoyancy because the tail is denser than the head. *D* is the drag on the body and fins and *T* is the forward thrust produced by the swimming movements.

If the fish is swimming horizontally at a constant velocity, it must be in equilibrium. The thrust T must equal the drag D and

$$A + B + C = W$$

We will now take moments about the centre of gravity. If the distances from it of the lines of action of A, B and C are a, b and c, respectively,

$$Aa + Bb = Cc$$

These equations can be used to estimate the upward forces B and C needed for equilibrium by particular selachians (Alexander, 1965). A, W, a, b and c can all be measured or estimated. The equations can then be solved as a pair of simultaneous equations, giving values for B and C. In the case of a tope (*Galeorhinus*) weighing 3·5 kg, they gave values of 113 g wgt for B and 44 g wgt for C. Whenever the fish swims horizontally, whatever its speed, B and C must have these values.

The forces on a hydrofoil depend on its speed. If B is to be kept constant the lift coefficient of the pectoral fin must be reduced as the speed increases. There are muscles which could

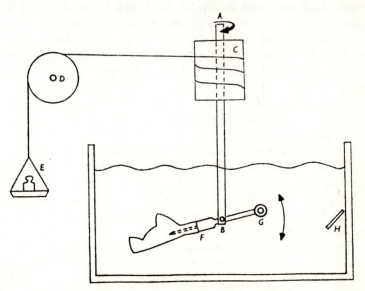

Figure 114. Apparatus used to investigate the forces which act on hetero-cercal tails. (From Alexander, 1965)

do this, either by reducing the angle of attack or by reducing the camber. If C is to be kept constant the lift coefficient of the tail must be reduced as the fish swims faster and moves its tail more rapidly from side to side.

Fig. 114 shows apparatus which was used to measure the vertical forces produced by heterocercal tails (Alexander, 1965). A tail is fixed upside down to one end of the bar FG, which is pivoted at B to a vertical shaft, suspended from a ball race at A. The shaft can be made to rotate by allowing the weights at E to fall. The tail is immersed in water in a large sink and moves through the water when the shaft rotates. It produces a down-ward force (since it is upside down) which can be measured by balancing it against weights bolted on at G. A chosen weight was bolted on (in addition to the weights needed to balance the weight of the tail) and the rate of rotation was adjusted so that the tail neither rose nor sank.

The results of the experiments were used to draw graphs of vertical force against speed (Fig. 115). Different parts of the tail were different distances from the shaft and moved at different speeds, and the speed shown on the graph is the speed of the point on the tail where the resultant force was judged to act. For both tope and dogfish (*Scyliorhinus*), the vertical force is roughly proportional to (speed)$^{1.5}$. If the tail were a rigid hyro-foil it would be proportional to (speed)^2. The difference must be due to the tail bending, so that its lift coefficient falls as the speed increases. One would expect this to happen, for the greater the speed the greater the forces tending to bend the tail and reduce the angle of attack α (Fig. 113b).

Though the lift coefficient falls as the speed increases it does not by any means fall fast enough to keep the lift constant. The arrows in Fig. 115 indicate the estimated upward force which the tail of each fish would have to produce, when it was swimming horizontally. The tope tail produced the right force at 38 cm/s and the dogfish tails at 47 and 68 cm/s.

The tail has a side-to-side movement superimposed on the forward movement of the fish. The speeds we have been dis-cussing refer to the side-to-side component but it seems from cinematograph films that this is about equal to the forward speed of the body. We may conclude that the tope tail would

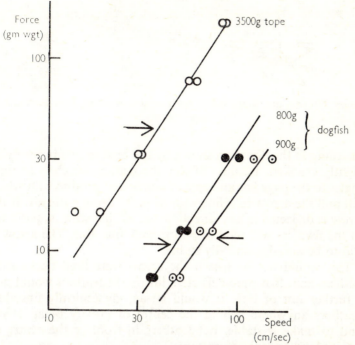

Figure 115. Graphs of the vertical force acting on tope (*Galeorhinus*) and dogfish (*Scyliorhinus*) tails, against the speed of movement in the apparatus of Fig. 114. The data for dogfish are not those originally published, which were obtained in experiments with tails which had been stored frozen (Alexander, 1965), but are from new experiments with fresh tails. The data for tope also refer to a fresh tail which had not been frozen. The arrows are explained in the text

have given the right upward force without muscular adjustment when the fish was swimming at about 38 cm/s, and that the dogfish tails would have given the right force at some speed around 60 cm/s. If the fish swam horizontally at any other speed the muscles of the tail fin would have had to have been used to alter the lift coefficient of the tail and adjust the force to the right value. Dogfish often swim at about 30 cm/s in aquaria, but they can swim a lot faster.

Aerodynamic stability

Fig. 116 shows an arrow with feathers at its rear end. It is

Figure 116. An arrow travelling horizontally, showing how the feathers give stability

travelling in the direction shown by the broken line but is flying slightly crooked. Because of this the feather which is at right angles to the page has an angle of attack and produces lift which will pull the arrow into line with its path through the air. If the arrow is deflected in any direction out of line with its path, lift on the feathers will tend to pull it back into line. The arrow is said to be aerodynamically stable.

This would not be true if the arrow were fired backwards, feathers first. If it were deflected, lift on the feathers would pull if further out of line. It would be aerodynamically unstable. Feathers anywhere behind the centre of gravity of an arrow tend to make it stable, but feathers in front of the centre of gravity tend to make it unstable.

An object may be aerodynamically or hydrodynamically stable for one sort of deflection, but unstable for another. Fig. 117 shows the names given to different sorts of deflection. Rotation about a vertical axis is called yaw and rotation about a transverse one is called pitch. The arrow shown in Fig. 116 should be stable both in yaw and in pitch.

Rolling is rotation about a longitudinal axis. The feathers on an arrow resist rolling but they have no tendency to correct a roll, once it has occurred. An arrow is neither stable nor unstable in roll.

Stability of a shark

Mustelus is a small shark (Fig. 117). Harris (1936) used a wind tunnel to investigate its stability. He had to use a cast of the fish rather than the fish itself, because a dead fish could not be kept in a natural posture in a wind tunnel, and it would flap like a flag. The cast was carefully made from a freshly-

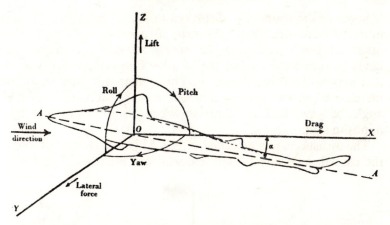

Figure 117. The cast of *Mustelus* used for the investigation of stability, in a diagram showing the meanings of "pitch", "roll" and "yaw". (From Harris, 1936)

killed fish 90 cm long. The body and all the fins were kept straight. The body was cast in plaster but the fins were cast in wax so that they could be detached. The plaster was painted with shellac to make it smooth.

Living fish sometimes glide through the water with their bodies more or less straight, after a spell of active swimming. The cast could only be used to investigate stability in this situation, for it could not imitate swimming movements.

The cast was fixed in the wind tunnel, attached to an aerodynamic balance in various ways depending on which forces were to be measured. A wind of 18 m/s was used. Since the kinematic viscosity of air is 14 times the kinematic viscosity of water (Table 4, page 210), this gave the same Reynolds number as a water speed of 130 cm/s. A 90 cm *Mustelus* could surely swim at this speed.

The most interesting of Harris' results concern stability in yaw. The cast was arranged horizontally in the wind tunnel and measurements were made with it facing straight into the wind and at various angles of yaw. Among other quantities, the yawing moment was measured. This is the couple tending to rotate the fish in the horizontal plane (i.e. to alter the angle of yaw).

Some of the results are displayed in Figs. 118 and 119. Yaw in one direction is taken arbitrarily as positive and yaw in the other is taken as negative. The yawing moment is positive if it tends to rotate the cast in the positive direction, and negative if it acts in the other direction. This means that yaw tends to increase when the yawing moment has the same sign as the angle of yaw, and to diminish when the moment has the opposite sign.

The measurements recorded in Fig. 118 were made with only the caudal, anal and posterior dorsal fins attached. These fins

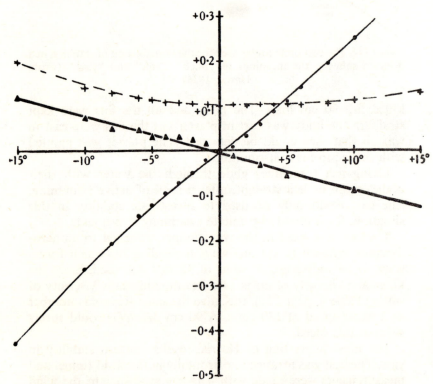

Figure 118. The results of an experiment in a wind tunnel with the cast shown in Fig. 117 without pectoral, pelvic or anterior dorsal fins. Various quantities are plotted against the angle of yaw. The thin broken line gives the drag in lb wgt, the thin continuous line the lateral force in lb wgt and the thick line the yawing moment in in lb wgt × 1/10. (From Harris, 1936)

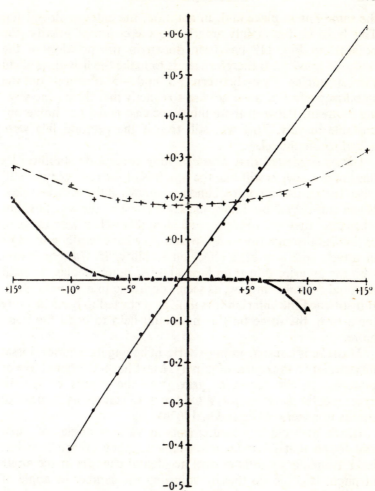

Figure 119. As Fig. 118, but with only the pectoral and pelvic fins missing.
(From Harris, 1936)

are all vertical and all lie well posterior to the centre of gravity, so they work like the feathers of an arrow. When the angle of yaw is positive the yawing moment is negative and *vice-versa*, so the yawing moment always tends to correct the yaw. The cast is stable in yaw.

Fig. 119 shows another set of measurements, obtained with

the same fins in place and, in addition, the anterior dorsal fin. This large fin lies mainly anterior to the centre of gravity (the point O in Fig. 117 has been drawn in the position of the centre of gravity). It therefore tends to make the fish unstable in yaw. At angles of yaw between $+5°$ and $-5°$ it cancels out the stabilizing effect of the other fins so exactly that there is no yawing moment. Between these limits the cast is neither stable nor unstable in yaw. This was still true if the pectoral fins were added to the model.

Harris suggested that *Mustelus* may control its stability. Its fins are not always stiff like the wax fins of the cast, but are stiff when their muscles contract and go limp when the muscles relax. When stiff they should have the same effect as the wax fins, but when limp they will not: they will align themselves with the path of the fish through the water so that they have hardly any angle of attack and very little effect on stability. If *Mustelus* swam with the anterior dorsal fin limp and the other median fins stiff it would be stable in yaw, as seems desirable in a straight glide. If it stiffened the anterior dorsal fin as it started to bend its body for a turn, the force on this fin would help to bend the body more.

Mustelus is unusual among sharks in having its anterior dorsal fin anterior to the centre of gravity. Most have it either above or posterior to the centre of gravity so the results of Harris' experiments on *Mustelus* should not be taken as typical of sharks in general (Alexander, 1967b).

Harris also made measurements at various angles of pitch and found that *Mustelus* is unstable in pitch. When it swims level it must constantly correct accidental changes in the angle of pitch. It could do this by adjusting the camber or angle of attack of the pectoral fins.

Swimming by undulation

Fig. 120 was traced from a cinematograph film of an eel swimming. The nine outlines are numbered in order and they are superimposed as though the eel were making no progress, though the eel actually travelled forward. The body is thrown into waves which travel backwards. On the left side, for instance, outline 3 is labelled at the crest of a wave and this wave

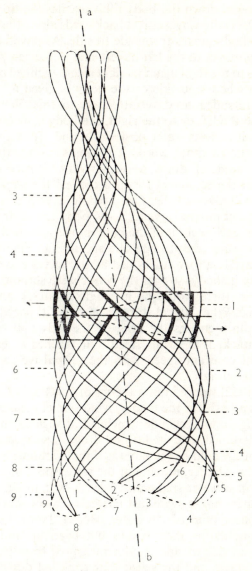

Figure 120. Outlines traced from successive frames of a cine film of a young eel (*Anguilla*) swimming, superimposed as if the eel were not moving forwards. (From Gray, 1933)

can be followed down the body till it reaches the tip of the tail in outline 9. As the waves travel backwards along the body each part of the body moves from side to side. As any short section of the body moves to the left its left side is inclined posteriorly. As it moves to the right again its right side is inclined posteriorly. The region where one side of the body is given a thick black outline is intended to draw attention to this. Whether it is travelling to the left or to the right, the body is inclined in such a way that it deflects water posteriorly and tends to drive the fish forwards. In other words it moves sideways through the water at an angle of attack, and a forward lift acts on it.

The faster the eel moves forward the less the angle of attack will be. It will always be less than would appear from Fig. 120 because the eel moves forward as its body moves from side to side, so that each section of the body moves obliquely forwards through the water, first to one side and then to the other. If the eel were travelling forward as fast as the waves travelled backwards along its body, so that the waves were stationary relative to the water, there would be no angle of attack and no propulsive force. When an eel starts swimming its speed increases, and this both increases the drag on its body and reduces the angle of attack. If it swims steadily it reaches a steady state, travelling at constant speed with the propulsive force exactly balancing the drag.

Taylor (1952) worked out a theory of this kind of swimming. He used as a basis for the theory, wind tunnel measurements of lift and drag on long cylinders inclined at various angles to the wind. He assumed that the lift and drag on each section of the body of an eel would be the same as if that section were part of a long straight cylinder, moving at the same speed and angle. He used films of a water snake and a leech swimming, to test the theory. Water snakes swim just like eels, except that they often keep their heads out of the water (Hertel, 1966). Leeches also swim by undulating their bodies but bend up and down instead of from side to side. Taylor measured from the films the size of the waves and the speed they travelled down the body. He used his theory to calculate the speed the animal should move through the water. The answers came out about right.

The propulsive force developed by a swimming eel depends

on the amount of backward momentum which the eel gives the water as it passes through. As with propellers (page 223) a given force can be obtained more economically by accelerating a lot of water to a low velocity than by accelerating less water to a higher velocity. To accelerate large quantities of water the eel must make undulations of large amplitude so that the band of accelerated water it leaves behind it is reasonably wide. It is not necessary for the whole body to make these large movements: indeed, it is better for the amplitude to increase gradually from a small value at the head to a large one at the tail (Lighthill, 1960). The amplitude increases along the body in eels (Fig. 120). It does so even more markedly in more ordinary teleosts which look as if they swim simply by wagging their tails from side to side but can be shown by cinematography to move their bodies in essentially the same way as eels (Gray, 1933). A tall tail fin helps to make swimming efficient, by making the band of accelerated water deep.

Let U be the velocity of a fish relative to the water and u the velocity of the waves along its body. A large amplitude and a tall tail fin would help to make U large, for a given value of u. Lighthill (1960) calculated that the theoretical efficiency would equal $(u + U)/2u$. Gray (1933) measured U and u from films of various teleosts and found that U varied from about $0 \cdot 5u$ to $0 \cdot 7u$, indicating efficiencies between $0 \cdot 75$ and $0 \cdot 85$. The film of a leech which was used by Taylor (1952) shows $U = 0 \cdot 2u$ and indicates an efficiency of $0 \cdot 6$. Notice that Lighthill's expression for efficiency can never give values less than $0 \cdot 5$.

Errant polychaetes like the ragworm, *Nereis*, have parapodia projecting from their sides. Fig. 121 shows a polychaete swimming. Waves are travelling towards the right. A parapodium on the outside of a bend is inclined to the right before the crest of a wave has reached it, and to the left after the crest has passed. It swings to the left as the wave passes and acts as a paddle, tending to drive the worm to the right. This would be true even if the parapodia were fixed at right angles to the body, as in the figure. In fact, they are moved in such a way as to enhance the paddling action (Clark, 1964). When *Nereis* and similar worms swim, they travel in the same direction as the waves along their bodies. They need forward waves to drive

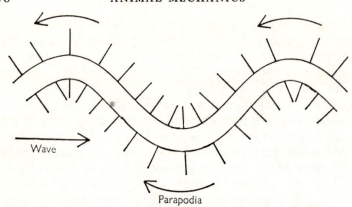

Figure 121. A diagram showing how waves travelling anteriorly along the body of a worm with parapodia can propel the worm forwards

themselves forwards, while eels need backward waves. Taylor (1952) included their sort of swimming in his mathematical analysis.

Spermatozoa swim by eel-like movements or by slightly more complicated helical waves (Holwill, 1966). Like eels, they are driven forwards by waves which travel backwards, but this cannot be explained in the same way as for eels. A spermatozoon has a very thin tail which moves slowly from side to side (the frequency of the waves may be high but their amplitude is very small), so the tail works at a very low Reynolds number, of the order of 10^{-3} or less. At such Reynolds numbers, lift is negligible and the movement of the animal must be explained entirely in terms of forces due to viscosity. This is not very easy to do.

CHAPTER 7

VIBRATIONS AND CHANGING FORCES

THE physical topics dealt with in this chapter are discussed more fully in textbooks on acoustics such as Wood (1940) and Randall (1951). Although the study of sound is probably not generally thought of as a branch of mechanics, it is included here because it is a mechanical phenomenon and because it is closely related to the other topics discussed in this chapter.

Free vibration
  Fig. 122a represents a hacksaw blade which is clamped at one end and has a weight of mass m attached to the other. If it is bent to one side and released it will vibrate from side to side. Consider it, vibrating, at a moment when the weight is a distance x to one side of its initial position (Fig. 122b). If the stiffness of the blade is S there will be an elastic restoring force $-Sx$, tending to drive the weight back to its initial position. The minus sign indicates that the restoring force acts in the direction opposite to the displacement. This force will give the weight an acceleration a in accordance with Newton's second law of motion (equation 1, page 1), so that if the mass of the blade is small enough to be ignored

$$-Sx = ma \tag{40}$$

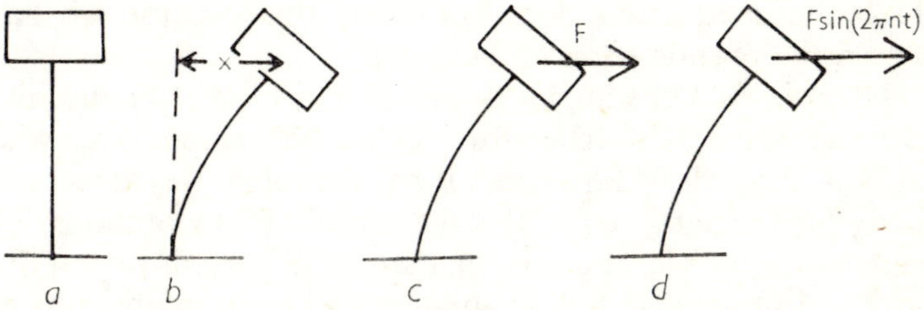

Figure 122. These diagrams are explained in the text

The course of the vibration can be worked out from this equation, using calculus. It can be shown that if the weight was displaced by a distance x_0 and released at a moment which is taken as zero time, the value of x at time t will be given by

$$x = x_0 \cos (\sqrt{[S/m]} . t) \tag{41}$$

The value of the cosine fluctuates between $+1$ and -1 as time passes, so x fluctuates between $+x_0$ and $-x_0$ (Fig. 123a). In other words, x_0 is the amplitude of the vibration. The vibra-

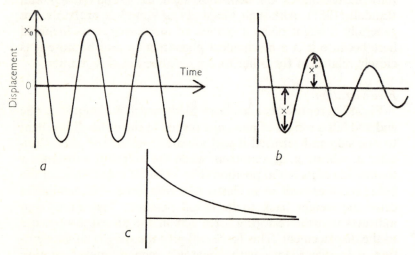

Figure 123. Graphs of displacement against time after release at an initial displacement x_0 of (*a*) an undamped oscillatory system, (*b*) a moderately damped system, and (*c*) a heavily damped system

tions of a real hacksaw blade would of course die away gradually but our equation does not show this because we have ignored the phenomenon of damping.

The cosines of the angles 0 rad, 2π rad ($360°$), 4π rad, etc., have the value $+1$. Therefore x has the value $+x_0$ when $(\sqrt{[S/m]} . t)$ equals 0, 2π, 4π, etc. It has the value $+x_0$ at intervals of $2\pi\sqrt{[m/S]}$, and $2\pi\sqrt{[m/S]}$ is the period of its vibrations. The frequency of the vibrations is the number of periods in unit time and is $\sqrt{[S/m]}/2\pi$. If S is in dyn/cm and m in grams, this expression gives the frequency in cycles per second.

Whenever the blade is pulled to one side and released it will vibrate with this frequency. It will also vibrate with this frequency if it or the clamp holding it is given a knock. Notice that the frequency depends on the ratio of S to m. The stiffer the blade the higher the frequency, and the heavier the weight the lower the frequency.

The vibrations of a real hacksaw blade would die gradually away because the resilience of steel is less than 100% (page 70) and because energy would be used overcoming drag on the blade as it moved to and fro through the air. Vibrations that die away for reasons like these are described as damped vibrations.

Materials have resiliences less than 100% because they are viscous as well as elastic. The force needed to overcome the viscosity is proportional to the rate of shear in the material (page 85). In the case of the vibrating hacksaw blade this force at any instant will be proportional to the velocity v of the weight. The drag on each portion of the blade may be proportional to the velocity or to the square of the velocity, depending on the Reynolds number (page 220). Many of the vibrating systems zoologists are concerned with involve small objects vibrating with small amplitudes, so that their Reynolds numbers are low. We will assume that the Reynolds numbers are low enough for drag to be proportional to velocity. This makes our problem pleasantly simple. Both the forces responsible for damping—the one due to viscosity of the vibrating material and the one due to drag—are proportional to the velocity.

We can now write a more sophisticated version of equation (40), which takes account of damping. The force available to accelerate the weight is the elastic restoring force minus the force (proportional to the velocity of the weight) which is responsible for damping. If this latter force is Kv our new equation is

$$-Sx - Kv = ma$$

or
$$Sx + Kv + ma = 0 \qquad (42)$$

From this equation it is possible to derive a more sophisticated version of equation (41), which is

$$x = x_0 . e^{-Kt/2m} \cos (\sqrt{[S/m - K^2/4m^2]} . t) \qquad (43)$$

The exponential term $e^{-Kt/2m}$ equals 1 when $t = 0$, since $e^0 = 1$. It diminishes as t increases; it is $1/e$ ($= 0 \cdot 37$) when $t = 2m/K$, $1/e^2$ ($= 0 \cdot 14$) when $t = 4m/K$ and so on. Because of it, x has a lower value on each successive occasion when the cosine equals 1: the amplitude of the vibrations gets gradually less. This is illustrated by Fig. 123b. The smaller $2m/K$ is (i.e. the larger K/m is), the more rapidly the amplitude falls.

Damping affects the frequency of the vibrations, as well as making them die away. We worked out from equation (41) that the frequency of undamped vibration was $\sqrt{[S/m]}/2\pi$. In the same way equation (43) gives the frequency of damped vibrations as

$$\sqrt{[S/m - K^2/4m^2]}/2\pi$$

which diminishes as K increases.

The distances x' and x'' in Fig. 123b are the amplitudes of two successive peaks in the vibration, half a cycle apart. It can be shown that

$$\log_e(x'/x'') = K/4mf \tag{44}$$

where f is the frequency of vibration. The quantity $\log_e(x'/x'')$ is known as the logarithmic decrement of the vibration, and equation (44) can be used to calculate K from it.

In cases when most of the damping is due to the viscosity of the elastic material, and drag can be ignored, the ratio (x'/x'') can be used to calculate the resilience of the material, as explained on page 71.

It has been assumed so far that K is not too large. Equation (43) only applies when $K < 2\sqrt{[mS]}$. If $K > 2\sqrt{[mS]}$ the system is too heavily damped to vibrate. If our weighted hacksaw blade were immersed in treacle, it would not vibrate when it was bent and released. The damping would be so great that it would simply straighten gradually (Fig. 123c). The greater the damping, the more slowly it would straighten.

Now suppose that the hacksaw blade is to be used as an instrument for measuring forces. A horizontal force F is applied to its end, and bends it (Fig. 122c). The blade eventually settles with its end displaced by a distance F/S. The time it takes to settle will depend on the damping. If $K \ll 2\sqrt{[mS]}$ the blade will bend rapidly when the force is applied, but it will vibrate for some time before settling down in its final position. If $K \gg 2\sqrt{[mS]}$ the blade will bend slowly and take a long time to reach its final position. The blade will settle down soonest in its final position if $K = 2\sqrt{[mS]}$. When K has this value, which is just enough to prevent vibration, the system is described as critically damped. The blade will work best as a device for measuring forces if it is critically damped because critical damping makes it possible to take a reading at the earliest possible moment after the force is applied.

The degree of damping is not the only thing that affects the time the blade takes to settle down. It can be shown that the time taken by a critically damped system to move 90% of the way to its final position is about $4\sqrt{[m/S]}$, and the time taken to travel 99% of the way is $6\cdot7\sqrt{[m/S]}$ (this is about equal to the period of vibration the system would have if it were not damped). A large value of m/S makes an instrument sluggish. The damped hacksaw blade would work best as an instrument for measuring forces, if it had no weight at its end. This would not of course make $m = 0$, for it would then be necessary to take account of the mass of the blade itself, which we have ignored.

Zoologists often want records of forces that change rapidly. This can only be done with instruments that settle down after a force has been applied, in a time that is short compared to the duration of the force. An instrument that takes 1 s to settle down is no use for measuring a force that only lasts for $0\cdot1$ s. Machin (in Donaldson, 1958) has described the properties of various instruments which can be used for measuring forces.

Vibrations of spiders' webs

Spiders rest on their webs, waiting for prey to be caught. They have sense organs in their legs which are sensitive to vibrations and they seem to rely on vibrations of the web to detect prey. They can be trained to distinguish between different frequencies when their webs are made to vibrate artificially.

Spiders' webs are elastic. Insects have mass. A spider's web with an insect caught in it has a natural frequency of vibration, which depends on the mass of the insect and the stiffness and damping of the web. If the insect collides with the web and sticks to it, or if, once caught, it struggles, it will set the web vibrating at this frequency.

Parry (1965) investigated the vibrations of webs with insects caught in them. He gave wooden rings to captive specimens of the spider *Tegenaria*, which built webs across the rings. The rings with webs on them were fixed in apparatus with a transducer touching the web. This transducer responded to vibrations by producing an electrical signal which was amplified and displayed on a cathode-ray oscilloscope.

Parry used small cockroaches, about 1 cm long, as "prey". He anaesthetized a cockroach and set it in the centre of a web, in the apparatus with the transducer. He recorded the vibrations in the web, when the cockroach recovered and started struggling. The vibrations were irregular and included components with various frequencies, but the component which reached the largest amplitude had a frequency of about 15 cycles/s (Fig. 124). Parry showed that this was the frequency of free vibration of the insect and web by putting an anaesthetized cockroach on the web and giving it a knock (by dropping a small bead on it) before it recovered from the anaesthetic. This started a series of damped vibrations in the web, with a frequency of 15 cycles/s.

Parry also detected vibrations with a frequency of about 100 cycles/s, in webs with struggling cockroaches. This proved to be another natural frequency of vibration of insect and web. The vibration at 15 cycles/s is simple vibration, to and fro, at right angles to the plane of the web. The vibration at 100 cycles/s involves rotation of the cockroach's body, first in one direction and then in the other.

It would be possible for spiders to judge the weights of insects in their webs from the frequencies of the vibrations set up by them. It is by no means certain that they do this. Indeed, Parry thinks that prey may not be recognized by their free

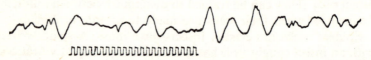

Figure 124. Typical vibrations in a spider's web, due to slight movements of an insect. Time marker (below): 50 cycles/s. (From Parry, 1965)

vibrations in the web, but by the very sudden jerks that occur in a web with an insect in it. These jerks may be due to threads in the web snapping back into position after being displaced. They can be reproduced by moving an insect leg very gently in the web.

Otoliths

All vertebrates have two or three otolith organs in each ear. Their structure is shown in Fig. 125*a*. The otolith itself consists mainly of calcium carbonate. It is contained in a fluid-filled cavity and one of its sides is attached to the wall of the cavity by flexible tissue which contains the sensitive processes of sensory cells. If the otolith moves and bends the sensitive processes, the sensory cells are excited. The sensory cells of an otolith organ in the position shown in Fig. 125*a* would be stimulated by horizontal movements of the otolith.

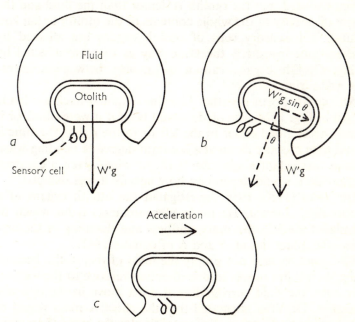

Figure 125. Diagrams showing the structure of an otolith organ and how it is affected by tilting and acceleration

Otoliths are much denser than water, on account of the calcium carbonate they contain. They are surrounded by fluid of about the same density as water. If the weight of an otolith is W and its specific gravity is ρ, its weight in the fluid surrounding it is $W' = W(\rho - 1)/\rho$ (see page 185). When the otolith organ is in the position shown in Fig. 125a this weight acts parallel to the sensory processes. When the organ is rotated through an angle (Fig. 125b) the weight has a component which acts at right angles to the sensory processes. It moves the otolith relative to the cavity and bends the sensory processes, as indicated in the figure. When the head is tilted, the frequency of action potentials changes in the sensory nerves of the otolith organs.

Accelerations also affect the otolith organs. If the otolith organ shown in Fig. 125c has an acceleration towards the right the dense otolith will tend to lag behind as shown. Of course, when the otolith lags behind the surrounding fluid must be driven forward, but the otolith is denser than the fluid and the centre of gravity of the whole contents of the otolith organ lags behind. The sensory cells of otolith organs are affected by accelerations in exactly the same way as they are affected by tilting. Otolith organs cannot distinguish between acceleration and tilting.

An otolith organ is rather like the weighted hacksaw blade shown in Fig. 122a. The blade would bend if it were tilted, or if it were accelerated either to the left or to the right. The time it would take to settle down at a constant degree of bending when held at a constant angle after a tilt or when given a constant acceleration, would depend on how heavily it was damped.

De Vries (1950, 1956) investigated the otolith organs of a teleost fish, *Acerina*. He made measurements from which he calculated the effective mass, stiffness and damping of the otoliths—the quantities m, S and K of equation (42).

The quantity m is not simply the mass of the otolith because when the otolith moves in one direction relative to the walls of its cavity, the fluid surrounding it must move in the opposite direction. De Vries suggested that the effective mass would be about equal to the mass of the otolith plus the mass of an equal volume of water. We need not trouble about the justification for this, because the difference between the actual mass and the effective mass is not large enough to make very much difference to our conclusions. We will be estimating S from measurements on the saccular otolith (the largest otolith) of a 21 g *Acerina*. This otolith must have weighed about 0·033 g. Its density would have been about 2·9 so its effective mass m must have been about $0·033 + (0·033/2·9) = 0·044$ g. Its weight in the fluid surrounding it (W' of Fig. 125) must have been about $0·033 - (0·033/2·9) = 0·022$ g.

Fig. 126 shows how de Vries measured S. The head of a freshly killed fish is clamped firmly in position over the photographic plate P. Head and plate can be rotated about the horizontal axle A which is exactly in line with an X-ray tube at F. When the head points upwards as shown each otolith is pulled posteriorly by a force $W'g$. If the axle is rotated through 180° so that the head points down, each otolith is pulled anteriorly by a force $W'g$. Turning the axle shifts the otoliths relative to the surrounding bones. De Vries took X-ray photographs of the head in both positions and measured the movements of the otoliths.

He found that the saccular otolith of a 21 g *Acerina* shifted 0·014 cm in an experiment of this sort. A forward force $W'g$ shifted the otolith 0·007 cm forward from its middle position and a backward force $W'g$ shifted it 0·007 cm back. We have estimated that W' must have been about 0·022 g so $W'g$ must

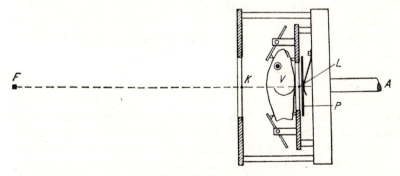

Figure 126. Apparatus used to investigate the mechanical properties of fish otoliths. (From de Vries, 1950)

have been about 22 dyn and the stiffness must have been about $22/0 \cdot 007 = 3,000$ dyn/cm. Other otoliths were rather less stiff.

The otolith would settle down soonest in its new position when the fish changed its posture, if it were critically damped. For this, K would have to equal $2\sqrt{[mS]} = 2\sqrt{[0 \cdot 044 \times 3,000]} = 23$ dyn s/cm. If it were critically damped, it would travel 99% of the way to its final position in $6 \cdot 7\sqrt{[m/S]} = 6 \cdot 7\sqrt{[0 \cdot 044/3,000]} = 0 \cdot 025$ s. The sacculus could provide very prompt information about changes of position.

De Vries estimated K from the results of experiments in which he made the fish's head vibrate. He attached the axle A (Fig. 126) to an eccentric driven by a motor, so that the whole assembly vibrated. He took X-ray photographs, as it vibrated. Since the fish's head and the photographic plate vibrated together, the image of the skull was sharp. Since vibration involves acceleration first in one direction and then in the other, forces acted on the otoliths first in one direction and then in the other and they moved relative to the skull. The images of the otoliths were blurred, but the ends of each blur were clear enough for the amplitude of vibration, relative to the skull, to be measured.

This is a case of forced vibration. We will deal presently with the theory of forced vibrations and we will see that there is a range of frequencies (around the resonant frequency) in which the amplitude of forced vibrations depends largely on K (page 291). De Vries carried out his experiments in this range of frequencies and estimated that for longitudinal vibrations of the saccular otolith of *Acerina*, $K \rightleftharpoons 15$ dyn s/cm. This is only a little less than the ideal value which would give critical damping.

Semicircular canals

Nearly all vertebrates have three semicircular canals, set mutually at right angles, in each ear. They are filled with the same fluid as the rest of the inner ear. Each opens at both ends into one of the cavities of the middle ear and has a swollen ampulla at one end (Fig. 127). The ampulla is almost completely blocked by a gelatinous structure known as the cupula, which is attached at its base to the wall of the ampulla and can move like a swing door. There are sensory cells at the base of the

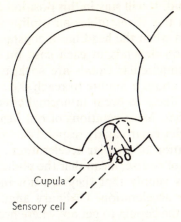

Figure 127. A diagram of a semicircular canal. The broken line shows how the cupula would be displaced relative to the canal by clockwise rotation

cupula, just like the sensory cells of otolith organs. The frequency of action potentials in the sensory nerve of a semicircular canal increases if the cupula is made to swing in one direction and decreases if it is made to swing in the other.

The semicircular canals are sensitive to rotation. When the head starts rotating the fluid in some or all of the canals (depending on the axis of rotation) tends to lag behind. If the canal shown in Fig. 127 were rotated clockwise the fluid would lag behind as it accelerated, so the fluid would move anticlockwise relative to the walls of the canal and the cupula would be deflected to the right as indicated by the broken outline.

The fluid in a semicircular canal has mass. The attachment of the cupula to the wall of the ampulla is elastic. Movements of the fluid relative to the wall are damped by the viscosity of the fluid. De Vries (1956) has estimated the mass, stiffness and damping for human semicircular canals. It seems that the damping is about 20 times the critical value.

If the head is rotated faster and faster with a constant angular acceleration, the cupula will be deflected more and more until it reaches a position, relative to the wall of the canal, at which the elastic restoring force of the cupula provides the force needed to accelerate the fluid. It will stay in this position until the acceleration changes. In this it resembles an otolith which settles in a constant position when the head has a constant linear acceleration. The position depends in each case on the acceleration. However, the semicircular canals are so heavily damped that they would take about a minute to reach the constant position. No acceleration likely to occur in normal circumstances would last as long as that. Most rotations of our heads take a second or less. The semicircular canals respond far too slowly to give any direct indication of angular acceleration.

This does not, of course, mean that the semicircular canals are useless. It means simply that they are not instruments which measure angular acceleration. Ordinary angular accelerations are too brief for the cupula to get anywhere near its equilibrium position, so the elastic restoring force of the cupula can be ignored. The movement of fluid relative to the walls of the canal is controlled almost entirely by the viscosity. This means that it is not the displacement of the fluid and cupula that is proportional to the acceleration, but the velocity. Suppose the head rotates with a constant angular acceleration α which gives the fluid a velocity $k\alpha$ relative to the walls of a canal. Suppose this continues for a time t which is not more than a few seconds. The angular velocity of the head will reach a value αt. The fluid will move a distance $k\alpha t$ in the canal. The distance the fluid moves and the amount the cupula is displaced at any instant are proportional to the angular velocity. The semicircular canals are not instruments for measuring angular acceleration, but instruments for measuring angular velocity.

We assumed constant angular acceleration in the last paragraph for the sake of simplicity, but it was not necessary to assume it. The displacement of the cupula is proportional to the angular velocity at the same instant, throughout any brief sequence of angular accelerations and decelerations.

Forced vibration

We return to the weighted hacksaw blade of earlier pages. Suppose that it is acted on by a force $F \sin (2\pi n t)$, as shown in Fig. 122d. This force is a fluctuating one. It fluctuates between $+F$ and $-F$ (that is, between F in one direction and F in the opposite direction). The frequency of its cycles is n and it makes the blade vibrate at this frequency.

Equation (42) (page 281) applies to free vibration. It can be modified to apply to forced vibration, by inserting the fluctuating force.

$$Sx + Kv + ma = F \sin (2\pi nt) \qquad (45)$$

From this equation it can be shown that the forced vibrations will settle down with a steady amplitude A, whose value is given by the equation

$$A = F/\sqrt{[(S - 4\pi^2 n^2 m)^2 + (2\pi nK)^2]} \qquad (46)$$

A is largest when the square root is least, which is the case when the frequency is

$$\sqrt{[S/m - K^2/2m^2]}/2\pi$$

This is lower than the frequency of free vibrations (page 282). It is called the resonant frequency of the system. At this frequency equation (46) becomes

$$A = F/K\sqrt{[S/m - K^2/4m^2]}$$
$$= F/2\pi fK$$

where f is the frequency of free vibrations. Notice that K is important in this equation. The amplitude depends largely on the damping, in forced vibrations at the resonant frequency.

The damping affects the amplitude much less at other frequencies. When n is much less than the resonant frequency the terms which include n in equation (46) can be ignored, so that the equation can be written

$$A \simeq F/S$$

The amplitude depends on the stiffness and hardly at all on the damping. On the other hand, when n is much larger than the resonant frequency the only part of the denominator in equation (46) that need be considered is the term including the highest power of n, and

$$A \simeq F/4\pi^2 n^2 m$$

Notice that m figures in this approximate equation, but K and S do not. At frequencies well below the resonant frequency, the ratio of A to F depends mainly on the stiffness of the system. At frequencies near the resonant frequency, it depends largely on the damping. At frequencies well above the resonant frequency, it depends mainly on the mass and the frequency.

Forced vibrations happen at the same frequency as the fluctuating force that causes them, but out of phase with it. If they were in phase with it, so that a force $F \sin (2\pi nt)$ caused displacements $A \sin (2\pi nt)$, a graph of force against displacement would be a straight line. In fact, the vibrations lag behind the force so that the displacement is given by the expression $A \sin (2\pi nt - \delta)$, and a graph of force against displacement forms a loop. This is a hysteresis loop (see page 70) and it shows that work is being done by the force. The area of the loop gives the work done in each cycle.

Work done against the stiffness of a vibrating system is recovered in the elastic recoil. Work done accelerating the system is recovered when it decelerates. The only work that is lost in a complete cycle of movements is the work done against damping. The power needed to maintain the vibrations at a constant amplitude is this amount of work, multiplied by the frequency. It can be shown that the power is $2\pi^2 n^2 A^2 K$. An increase in the damping coefficient K will obviously increase the power needed to maintain vibrations at a given amplitude.

The phase lag δ between the vibration and the force is given by the equation

$$\tan \delta = 2\pi nK/(S - 4\pi^2 n^2 m) \qquad (47)$$

The greater the damping, the greater the lag.

The viscoelastic properties of materials are often studied by fixing samples in apparatus in which they are stretched or sheared by a sinusoidal force. The phase lag is measured and $\tan \delta$ is quoted as the loss tangent of the material. Its value depends on the frequency. An equation relating it to the resilience of the material has already been given (equation 9, page 71).

A vibrating body has its greatest velocity when it is passing through its mean position. If the vibration is sinusoidal this velocity is $2\pi nA$, where n is the frequency and A the amplitude.

$2\pi nA$ is the velocity amplitude. The ratio of the amplitude of the force applied to a system to set up forced vibrations, to the velocity amplitude of the vibrations, is the mechanical impedance of the system (strictly, it is the modulus of the mechanical impedance). If the mechanical impedance is Z and the amplitude of the force F

$$Z = F/2\pi nA$$

and from equation (46)

$$Z = \sqrt{[(S - 4\pi^2 n^2 m)^2 + (2\pi nK)^2]}/2\pi n \qquad (48)$$

Insect flight muscles

Every contraction of an ordinary striated muscle fibre is started off by a nerve impulse, so the nerve impulses arriving at a fibre are at least as numerous as the contractions it makes. Insect flight muscles are different. A blowfly may beat its wings 120 times per second although only 3 nerve impulses per second are arriving at each fibre. The nerve impulses apparently put the muscle into a state of oscillatory contraction whose frequency bears no relation to the frequency of the impulses. However, the oscillatory contraction is only possible if the muscle has a suitable mechanical load to drive. If a piece of muscle is detached from the wings it cannot be made to oscillate unless a suitable alternative load is supplied (Pringle, 1967).

Machin and Pringle (1959) studied the properties of the flight muscles of large beetles by making them drive artificial loads. The size of the beetles, their low wingbeat frequency and the anatomical arrangement of their muscles made them peculiarly suitable for the experiments. Most of the experiments were done on the Indian rhinoceros beetle, *Oryctes*.

Their apparatus is shown diagrammatically in Fig. 128. The body of the beetle is fixed firmly in a special holder. A thin wire is attached to the muscle at one end and to a moving coil vibrator at the other. When current is passed through the coil of the vibrator, it exerts a force on the muscle. Forces acting on the muscle are measured by a piezo-electric force transducer and changes in the length of the muscle are measured by means of the vane, mounted on the wire. The vane partly blocks the beam

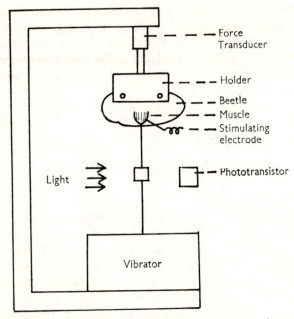

Figure 128. Apparatus used by Machin and Pringle (1959) to investigate the properties of beetle flight muscles

of light falling on the phototransistor and when the vane moves the electrical output of the phototransistor changes. The outputs of the piezo-electric transducer and the phototransistor are fed to the X and Y plates of a cathode ray oscilloscope, so that one coordinate of the spot on the oscilloscope screen gives the tensile force in the muscle, and the other gives its length.

The vibrator can be used in two different ways. It can be used to extend the muscle to a particular length and hold it there while the force is measured. The sloping lines which appear in all the graphs in Fig. 129 show force, measured in this way, plotted against length. The lower line was obtained with the muscle at rest and the upper one with the muscle stimulated electrically. The circuit operating the vibrator was so designed that the vibrator was held rigidly at each position at which it was set, and the muscle was prevented from making oscillatory contractions.

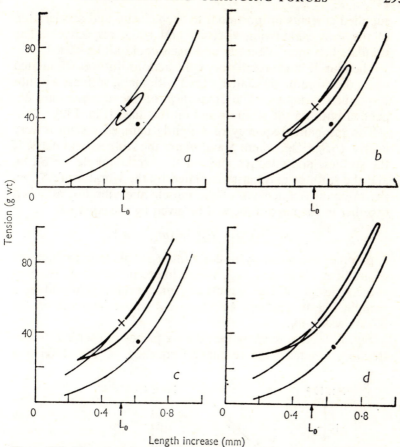

Figure 129. These graphs are explained in the text. (From Machin and
Pringle, 1959)

The vibrator could be connected to another circuit which
made it behave like a mechanical system with stiffness, viscosity
and mass. The vibrator exerts a force proportional to the current
supplied to it. Part of the circuit connected the phototransistor
to the vibrator in such a way as to produce a current proportional to the displacement of the moving vane from its
mean position: it made the vibrator exert a force proportional
to the displacement, just as if it was a spring or some other
mechanical system with stiffness. Other parts of the circuit

supplied currents proportional to the velocity and acceleration of the vane, simulating viscosity and mass, respectively. The simulated stiffness, viscosity and mass could all be varied.

The muscle made oscillatory contractions if it was stimulated while the vibrator simulated a load with mass, stiffness and not too much damping. The frequency of the contractions depended both on the stiffness and on the mass (Fig. 130).

The resonant frequency of a lightly damped system is very nearly $\sqrt{[S/m]}/2\pi$, where S and m are the stiffness and mass. If the stiffness and effective mass of the beetle muscle are S_1, m_1 and the stiffness and mass simulated by the vibrator are S_2, m_2 the resonant frequency n of the muscle and vibrator, attached together in the apparatus, will be given by the equation

$$n^2 \simeq (S_1 + S_2)/4\pi^2(m_1 + m_2)$$

A graph of n^2 against S_2 for constant m_2 would be a straight line of gradient $1/4\pi^2(m_1 + m_2)$, whose intercept on the X-axis was $-S_1$. A graph of $1/n^2$ against m_2 for constant S_2 would be a straight line of gradient $4\pi^2/(S_1 + S_2)$ whose intercept on the X-axis was $-m_1$.

Fig. 130 seems to show that the frequency of oscillatory contraction, f, equals the resonant frequency, n. Fig. 130a is a

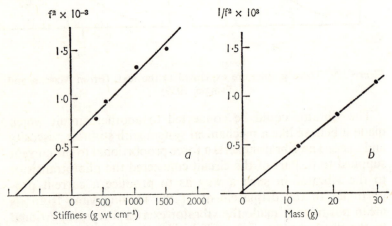

Figure 130. These graphs are explained in the text. (From Machin and Pringle, 1959)

graph of f^2 against stiffness for an experiment in which m_2 was kept constant at 30 g. The points lie reasonably near the straight line, whose gradient would be about 1/1,500 if the stiffness were given in dyn/cm instead of g wgt/cm. Since m_2 was 30 g and the mass of the muscle a fraction of a gram, $(m_1 + m_2)$ was about 30 g and $1/4\pi^2(m_1 + m_2)$ was about 1/1,200. This is reasonably near the gradient found. The intercept on the X-axis is about $-8 \cdot 5 \times 10^5$ dyn/cm (-850 g wgt/cm) which is reasonably near the value of $-S_1$ obtained from the experiments in which the muscle was stimulated but prevented from oscillating.

Fig. 130b is a graph of $1/f^2$ against m_2 for an experiment in which S_2 was kept constant at 4×10^5 dyn/cm. The points lie on a straight line whose gradient is 4×10^{-5}. If we use the value of S_1 given by the intercept in Fig. 130a, $4\pi^2/(S_1 + S_2) = 4\pi^2/(12 \cdot 5 \times 10^5) = 3 \times 10^{-5}$, which is close to the gradient found. The intercept on the X-axis is very small, corresponding to the small value of m_1.

The graphs of tensile force against muscle length which are displayed on the oscilloscope screen during oscillatory contraction, are loops (Fig. 129). The spot describing the loop moves anticlockwise, showing that the tension is greater at any given length during contraction than during extension, and that the muscle is doing work. This contrasts with the situation represented in Fig. 30c where work is being done on a passive piece of material, and movement is clockwise.

The area of the loop obtained from the muscle indicates the amount of work done in each cycle of contraction and relaxation. Fig. 129 shows how it depends on the amount of damping. The area is greatest in Fig. 129c and less both when there is more damping (Fig. 129a, b) and when there is less (Fig. 129d). The muscle does little work when the damping is high because the amplitude (i.e. the distance through which it contracts) is then low. It does little external work when the damping is low because most of the work it does is then used overcoming its own viscosity. The power output can be calculated by multiplying the work per cycle by the frequency. The highest sustained value recorded by Machin and Pringle was equivalent to 30 W/kg muscle. They believe that much higher power outputs occur in flight.

Electrical analogues

Fig. 131 is a diagram of an electrical circuit. The potential difference between the terminals of the AC generator is V sin $(2\pi nt)$ at time t. The potential difference across the condenser which has capacity C is q/C when the charge in the condenser is q. The potential difference across the resistor whose resistance is R is Ri when a current i is flowing. The potential difference across the coil whose inductance is L is $L\alpha$ when the current is changing at a rate α. Hence

$$q/C + Ri + L\alpha = V \sin (2\pi nt)$$

This equation is very like equation (45) which describes forced vibrations of a mechanical system. The current is the rate at which the charge in the condenser is changing so the charge, current and rate of change of current are related to each other in the same way as the displacement, velocity and acceleration of a mechanical system. $1/C$ takes the place in the electrical equation of the stiffness S in the mechanical one. R takes the place of the damping coefficient K and L takes the place of the mass m. The voltage amplitude V takes the place of the force amplitude F. n is the frequency in both equations.

Hence, the electrical circuit shown in Fig. 131 is an analogue of a simple mechanical system such as the weighted hacksaw blade of Fig. 122d. If the circuit is built with $1/C$, R and L numerically equal to the stiffness, damping coefficient and mass of the mechanical system, it can be used to estimate the impedance of the mechanical system at any frequency. The current in the circuit will be numerically equal to the velocity of the

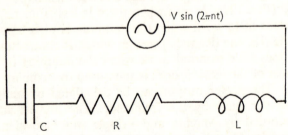

Figure 131. An electrical analogue of the mechanical system represented in Fig. 122(d)

mechanical system, if the AC voltage and the oscillating force have the same frequency and numerically equal amplitudes. The electrical impedance of the circuit, which is the voltage amplitude divided by the current amplitude, will be numerically equal to the mechnical impedance of the mechanical system.

There is usually little point in building an electrical analogue of a simple mechanical system, whose impedances can be calculated quickly. Electrical analogues can be useful in dealing with complicated systems, which would be tedious and difficult to analyse mathematically. Complex mechanical systems of course have correspondingly complex analogues, involving elements in parallel as well as elements in series. Machin (in Donaldson, 1958) has set out the rules for constructing electrical analogues.

The human ear is a very complicated mechanical system. The eardrum has mass and stiffness and its vibrations are damped. It is connected to the inner ear by three ossicles which have mass and whose movements are affected by stiffness and damping at their attachments. The air in the middle ear is alternately compressed and rarefied as the drum vibrates. It has stiffness in the sense that compression and rarefaction are resisted by the consequent pressure changes. The air also has mass and viscosity, and its effect on the impedance of the ear is complicated by the existence of chambers opening off the main cavity of the middle ear. The fluid in the inner ear has mass and viscosity and its vibrations involve alternate bulging of the oval and round windows, which have stiffness.

Several investigators have used electrical analogues of the ear. Zwislocki (1962), for instance, used an analogue to confirm his analysis of the factors contributing to the impedance of the ear. The impedances of complete and incomplete ears have been measured by various techniques involving measurement of the amplitude of vibrations caused by oscillating pressures. Zwislocki built an electrical analogue whose parts corresponded in number and arrangement to the main likely components of the impedance of the ear. He adjusted the capacities, resistances and inductances of the components, so that the electrical impedance of the circuit was close to the mechanical impedance of the ear at all frequencies, and the impedances of the appropriate parts of the circuit agreed with the impedances of incomplete ears. He then checked wherever possible that the capacities, resistances and inductances of individual components agreed reasonably well with the mechanical quantities they represented. For instance, he made sure that the inductance representing the mass of the tympanic membrane had a value close to the actual mass of the membrane. The fact that he was able to construct an analogue ot these exacting specifications is evidence that he had identified correctly the main factors contributing to the impedance of the ear.

The function of the middle ear is discussed later (page 309).

Sound

Fig. 132*a* represents a pulsating sphere—a sphere that is swelling and contracting periodically. We will suppose that it is surrounded by incompressible fluid. When the sphere swells it pushes fluid radially outwards in all directions. When it contracts, it draws fluid inwards. The pulsations of the sphere make the fluid vibrate. Suppose the sphere enlarges from a radius r to a slightly larger radius $(r + \delta r)$. Its volume increases by $4\pi r^2 \delta r$ (since the area of its surface is $4\pi r^2$) and a volume $4\pi r^2 \delta r$ of fluid is displaced. Consider a layer of fluid of radius s, which is concentric with the sphere. If this layer moves outwards by a distance δs the volume enclosed by it will increase by $4\pi s^2 \delta s$. This must equal $4\pi r^2 \delta r$, so

$$\delta s = (r^2/s^2)\delta r$$

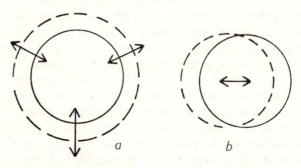

Figure 132. (*a*) Pulsation, and (*b*) vibration of a sphere

This means that the amplitude of the vibrations caused by the pulsations of the sphere is inversely proportional to the square of the distance from the centre of the sphere.

Fig. 132b represents a sphere which is vibrating from side to side. As it moves, it pushes the surrounding fluid out of the way. It makes the fluid vibrate as the pulsating sphere did, but in a more complicated way. The pulsating sphere made the fluid vibrate radially, and the amplitude was the same at the same distance in all directions. The vibrating sphere makes the fluid vibrate in such a way that fluid pushed from in front of it travels round to fill the space behind it. The amplitude is not the same at equal distances from the sphere in different directions. The amplitude in any given direction from the sphere is inversely proportional to the cube of the distance.

The disturbances which would be set up in incompressible fluids by pulsating or vibrating bodies are known as acoustic near fields. In real fluids, which are more or less compressible, another type of disturbance is set up as well. Disturbances of this second type are called acoustic far fields, or sound waves. Consider the pulsating sphere again. As it expands it compresses the fluid immediately round it, which then expands and compresses the next layer of fluid, and so on. A phase of compression travels radially outwards through the fluid. It is followed by a phase of rarefaction, started by the next contraction of the sphere. Alternating phases of compression and rarefaction travel through the fluid. They are sound waves. The particles of fluid vibrate backwards and forwards along the path of the sound. They move together where the fluid is compressed and apart where it is rarefied.

Sound waves travel at about $3 \cdot 3 \times 10^4$ cm/s in air and $1 \cdot 4 \times 10^5$ cm/s in water. If sound of frequency n travels with velocity c, one region of maximum compression will have travelled a distance c/n by the time the next one starts. Successive regions of maximum compression will be c/n apart and the wavelength of the sound is c/n. For instance, the wavelength of middle C (256 cycles/s) is $3 \cdot 3 \times 10^4/256 = 130$ cm in air and $1 \cdot 4 \times 10^5/256 = 550$ cm in water.

Sound involves the transmission of energy through the medium. The intensity of a sound is the power transmitted through a unit area set at right angles to the sound. If the intensity of the acoustic far field is I at a distance s from the centre of a pulsating sphere, the total power transmitted through the layer of fluid of radius s is $4\pi s^2 I$. If energy is neither gained nor lost as the waves travel outward, I must be inversely proportional to s^2. In fact, a small amount of energy is lost as heat, but this can generally be ignored. The intensity of an acoustic far field can generally be taken as inversely proportional to the square of the distance from the source. This is true of sound radiated by vibrating bodies as well as by pulsating ones, provided one compares points at different distances in the same direction. A vibrating object does not transmit any sound at right angles to its direction of vibration, and the intensity is greatest, at any given distance, in line with the vibrations.

The intensity I of a sound is related to the amplitude A of the movements of the vibrating particles of the material it is travelling through, and to the amplitude p of the fluctuations of pressure. It can be shown that for sound waves of frequency n travelling with velocity c in material of density ρ

$$I = 2\pi^2 n^2 \rho c A^2 \tag{49}$$

and
$$I = p^2/2\rho c \tag{50}$$

A and p are each proportional to the square root of I. If I is inversely proportional to the square of distance from the source, A and p will be inversely proportional to the distance itself. It must be emphasized that equations (49) and (50) apply only to acoustic far fields.

We can now see why the near and far fields are so called. The amplitude of vibrations of particles of fluid due to the near field effect is inversely proportional to the square or cube of distance from the source. The amplitude due to the far field effect is inversely proportional to the distance itself. The near field falls off with distance far faster than the far field. Close to a source, the near field effect is responsible for most of the amplitude of vibration but further from it, the far field effect is responsible for most of the amplitude. There is of course an intermediate distance at which the two contribute equally to the amplitude. This is about $0 \cdot 2$ wavelengths from the source. At distances well below $0 \cdot 2$ wavelengths, only the near field need usually be considered. At distances well above it, only the far field is important (Harris, 1964).

The pressure amplitude p is related to the amplitude of vibration A in far field sound by the following equation, as is apparent from equations (49) and (50)

$$p = 2\pi n p c A \tag{51}$$

This equation should only be applied to conditions well away from the source of sound, where the near field can be ignored. Close to the source p is less for a given value of A than the equation predicts.

Most sorts of microphone are affected by the pressure changes involved in sound. Their electrical output is proportional to the pressure amplitude, for sound of any given frequency. It is sometimes important to remember that the amplitude of vibration A, close to a source, may be much greater than equation (51) and the electrical output of a microphone would suggest.

Differences in intensity of sound are often expressed in decibels. If the intensities of two sounds are I_1 and I_2, the difference can be given as $10 \log_{10} (I_2/I_1)$ dB. For instance, if I_2 were three times I_1 the difference would be $10 \log_{10} 3 = 5$ dB. If I_2 were 100 times I_1 the difference would be $10 \log_{10} 100 = 20$ dB. It follows from equation (50) that if the pressure amplitudes of the two sounds are p_1 and p_2, $I_2/I_1 = (p_2/p_1)^2$ and the difference in intensity is $10 \log_{10} (p_2/p_1)^2 = 20 \log_{10} (p_2/p_1)$ dB. Measurements of microphone outputs give p_2/p_1 not I_2/I_1.

The intensity of a sound is often given as so many decibels above or below some reference intensity. If it is higher than the reference intensity it is given as a positive number of decibels; if lower, as a negative number. The reference intensity is often described in terms of pressure, and when this is done the pressure given is not the pressure amplitude but the root mean square pressure. This is $1/\sqrt{2}$ times the pressure amplitude. The most usual reference level is $0 \cdot 0002$ dyn/cm^2, which is close to the threshold of human hearing. It corresponds to an intensity of 10^{-16} W/cm^2.

Hearing by fishes

When sound travels through water in which a fish is swimming the whole fish vibrates. The otoliths tend to lag behind as the vibrating fish accelerates, first to one side and then to the other, so the otolith organs are stimulated by the sound.

The sensitivity of fish to sound can be investigated, if they can be trained to perform some action in response to sound. They may for instance be trained to swim over a barrier which rises almost to the surface of the aquarium, to avoid an electric shock. The frequency and the intensity of the sound stimulus can be varied, so that the minimum intensity to which the fish will respond can be found for a range of frequencies. Conditioning experiments like this have always to be carefully designed, to make sure that the fish is really responding to the sound and not, for instance, to the experimenter's activities.

The obvious way to produce sound for experiments of this sort is with a waterproof loudspeaker submerged in the aquarium. This leads to difficulties. The otolith organs are sensitive to the accelerations involved in sound and acceleration is proportional, at any given frequency, to the amplitude of vibration, A. The sound in the aquarium would have to be measured by a submerged microphone which would not be sensitive to the amplitude of vibration, but to the pressure amplitude, p. If the acoustic near field could be neglected, equation (51) could be used to calculate A from p. It could only be neglected, if the distance from the loudspeaker to the fish was much more than $0 \cdot 2$ wavelengths. If the sensitivity of the fish to sound of 100 cycles/s were to be investigated, the aquarium would have to be much more than 280 cm long. This would probably be inconvenient. If the fish were too near the loudspeaker for the near field to be neglected, it would be very difficult to estimate the amplitude of vibration from the pressure amplitude.

At high frequencies it would be feasible to have the aquarium long enough for the near field to be negligible. However, the very conditions that make it possible to ignore the near field cause another difficulty. If the aquarium is not small compared to the wavelength, stationary waves will be apt to develop in the aquarium (see page 317). This could mean marked differences in the amplitude of vibration, between one part of the aquarium and another. It would be very hard to know exactly what amplitude the fish had been exposed to in an experiment, because this would depend so much on its position in the aquarium.

The near field difficulty can be overcome by using a loudspeaker in the air, some distance above the surface of the water, to produce the sound. The sound in the water has a much lower intensity than the sound in the air (page 308), but it is almost pure far field sound and the amplitude of vibration can be calculated from the pressure amplitude, measured with a submerged microphone. The stationary wave difficulty can be partly overcome by incorporating material that absorbs sound in the aquarium wall (see page 309 and, on practical details, Ritchie, 1965).

Fig. 133 shows the results of experiments on several species of teleost. In *Holocentrus*, *Carassius* and *Ictalurus* the swimbladder either extends close to the ear or is connected to the ear by the Weberian ossicles, which increase the sensitivity of the fish to

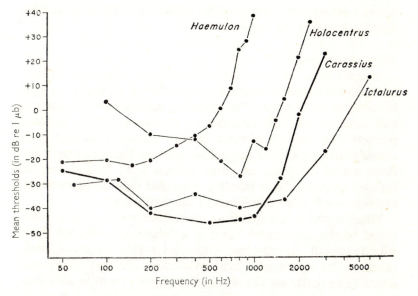

Figure 133. Graphs of auditory threshold against frequency for some teleost fishes. (From Jacobs and Tavolga, 1967)

sound (see page 320). Only in *Haemulon* are the otolith organs stimulated simply by the vibrations of the whole fish. Frequencies are given in herz (cycles/s) and the thresholds of hearing are given in decibels referred to a root mean square pressure of 1 μbar (1 dyn/cm², corresponding to a pressure amplitude of $\sqrt{2}$ dyn/cm²). At frequencies up to 200 cycles/s, *Haemulon* can just hear sounds of -20 dB. The corresponding pressure amplitude p can be worked out from this, since the intensity difference in decibels between sounds of pressure amplitudes p_1, p_2 is $20 \log_{10} (p_2/p_1)$

$$-20 = 20 \log_{10} (p/\sqrt{2})$$
$$p = \sqrt{2}/10 = 0\cdot14 \text{ dyn/cm}^2$$

The amplitude of vibration A can be worked out, using equation (51)

$$0\cdot14 = 2\pi n \rho c A$$
$$A = 0\cdot14/(2\pi n \times 1\cdot4 \times 10^5)$$
$$= 1\cdot6 \times 10^{-7}/n \text{ cm}$$

where n is the frequency. At 200 cycles/s this minimum audible amplitude is 8×10^{-10} cm.

Reflection of sound

Sound can be reflected and refracted just like light. Echoes are reflected sound. If sound is to be reflected regularly, as light is reflected by a mirror, the reflecting surface must be reasonably smooth. Any irregularities on it must be small, in comparison with the wavelength of the sound. A surface with irregularities which are large compared to the wavelength reflects sound diffusely, as paper reflects light. Sound can be focused by a parabolic reflector, provided that the diameter of the reflector is large compared to the wavelength.

When sound is reflected from an obstacle which is large compared to the wavelength there is a "shadow", shielded from sound, behind the obstacle. Obstacles which are small compared to the wavelength form no shadow and they scatter the sound. The scattered sound travels out from them in all directions.

Sound travels faster in water than in air, whereas light travels faster in air than in water. Sound is therefore refracted in the opposite direction to light, when it passes from air to water or from water to air. Sound cannot pass from air to water unless it strikes the water nearly vertically. If it has an angle of incidence more than 13°, it is totally reflected, just as light travelling in water is totally reflected if it strikes an air surface with an angle of incidence more than the critical angle. Notice that sound suffers total internal reflection in air whereas light suffers total internal reflection in water.

In conditions that do not result in total internal reflection, sound arriving at a boundary between two materials is partly reflected, and partly transmitted to the sound material. The proportion that is transmitted depends on the densities of the two materials, the velocities of sound in them and the angle of incidence. The largest proportion is transmitted if the angle of incidence is zero (i.e. if the sound strikes the surface at right angles). What follows assumes zero angle of incidence.

The pressure on either side of the boundary between two materials must be the same. If the pressure amplitude of the incident sound is p_i, that of the reflected sound p_r and that of the transmitted sound p_t, then, since p_i and p_r are on one side of the boundary and p_t on the other

$$p_i + p_r = p_t$$

p_r can be either positive or negative, as the reflected sound can be in or out of phase with the incident sound.

The amplitude of vibration on the two sides of the boundary must also be the same. If the amplitudes of the incident, reflected and transmitted sound are A_i, A_r and A_t, respectively

$$A_i + A_r = A_t$$

From these two equations one can work out the intensity of the transmitted sound. Let the intensities of the incident and transmitted sound be I_i and I_t, respectively. Let the densities of the two materials be ρ_1, ρ_2 and the velocities of sound in them c_1, c_2. Then, if the angle of incidence is zero it can be shown that

$$I_t/I_i = 4\rho_1 c_1 \rho_2 c_2/(\rho_1 c_1 + \rho_2 c_2)^2 \tag{52}$$

It can be shown that the product ρc for any material is equal to

the pressure amplitude of sound in it, divided by the corresponding velocity amplitude. The impedance of a system to forced
vibrations (page 293) is the amplitude of the applied force,
divided by the velocity amplitude. The quantities $\rho_1 c_1$, $\rho_2 c_2$ are,
in fact, impedances, each referring to a unit area at right angles
to the direction of the sound. They are called the characteristic
acoustic impedances of the materials. The density of air is
$1 \cdot 3 \times 10^{-3}$ g/cm³, and the velocity of sound in it is $3 \cdot 3 \times 10^4$
cm/s so the characteristic impedance of air is 43 dyn s/cm³. The
density of water is 1 g/cm³ and the velocity of sound in water
is $1 \cdot 4 \times 10^5$ cm/s so the characteristic impedance of water is
$1 \cdot 4 \times 10^5$ dyn s/cm³. By putting these values in equation (52)
we find

$$I_t/I_i = 4 \times 43 \times 1 \cdot 4 \times 10^5/(1 \cdot 4 \times 10^5)^2$$
$$= 1 \cdot 2 \times 10^{-3}$$

Thus, when sound travelling in air strikes a water surface at
right angles, only $0 \cdot 1\%$ of the sound energy enters the water.
Similarly, when sound travelling in water strikes an air surface
at right angles, only $0 \cdot 1\%$ of the energy escapes from the
water. In each case, $99 \cdot 9\%$ of the sound energy is reflected. An
air/water boundary is an extremely effective barrier to sound
energy.

It must be remembered that sounds of the same intensity in
air and water involve a much larger amplitude of vibration in
air, and a much larger pressure amplitude in water. This
follows from equation (51). When sound travels from air to
water, the intensity and the amplitude of vibration are greatly
reduced but the pressure amplitude is nearly doubled. $p_r \simeq p_i$ so
$p_t \simeq 2p_i$. When sound travels from water to air, the intensity
and the pressure amplitude are greatly reduced, but the amplitude of vibration is nearly doubled. The reflected sound is out
of phase with the incident sound (a compression is reflected as
a rarefaction) so $p_r \simeq - p_i$ and p_t is very small.

The phase of sound is reversed when it is reflected from a
boundary with a material of lower characteristic impedance
than the material it is travelling in, but not when it is reflected
from a boundary with a material of higher impedance.

When sound strikes a microphone, some of its energy is used

in driving the microphone and some is reflected. The proportions depend on the mechanical impedance of the microphone to vibrations at the frequency in question, and the area of its diaphragm. If a reasonably large proportion is to be used in driving the microphone, the impedance per unit area of the microphone must be close to the characteristic acoustic impedance of the air. This is often expressed by saying that the impedance of the microphone must be matched to the impedance of the air.

The invention of the phonograph preceded the invention of electronic amplifiers. The energy needed for indenting the record had all to be obtained from the sound. A large horn was used to collect sound energy from the large area of its mouth, and concentrate it on a small diaphragm at its apex. The horns used had the same shape as the ones used on early gramophones, like the one shown in the "His Master's Voice" trademark. Horns of this shape act as impedance matching devices. They can be made so that the impedance per unit area at the mouth is low and close to the characteristic impedance of air, while the impedance at the narrow end is high and close to the necessarily high impedance of the diaphragm. In this way energy collected at the large mouth of the horn is transmitted to the small diaphragm without too much loss.

Transmission and reflection are not the only possible fates of sound energy at a boundary. It may also be absorbed: that is, it may be converted into heat by being used to do work against viscosity. Porous materials absorb sound well, because of the viscous resistance to vibration of the contents of the pores. A thick layer of porous material absorbs more of the sound that strikes it than a thin layer, because the pores are longer.

The absorption coefficient of a surface is the fraction of the sound energy that strikes it, that is not reflected. So defined, it does not distinguish transmitted sound from absorbed sound. The coefficient of absorption of a wood block floor, for sound of 500 cycles/s, is about $0 \cdot 06$, but if the floor is covered by a thick carpet the coefficient may be raised to $0 \cdot 5$.

The middle ear

Terrestrial vertebrates live in air and the sounds that reach them are mostly sounds travelling in air, but the sensory cells used for hearing are in the liquid-filled inner ear. If a sound is to be heard it must set up forced vibrations in the fluid of the inner ear. The nature of these vibrations and the manner in which the ear detects them and distinguishes different frequencies are described in many other books and will not be described here (see, for instance, Littler, 1965).

The inner ear is enclosed by bone except at the oval and round windows where its wall is flexible. If the membrane of the oval window bulges out that of the round window must cave in and *vice versa*. The vibrations involved in hearing involve alternate bulging and caving of the membranes of the two windows. The membranes have elastic stiffness. The fluid that vibrates has mass and viscosity. Therefore the inner ear has a mechanical impedance.

This impedance can be calculated from the results of experiments by von Békésy (1949) on human ears. He applied oscillating pressures to the oval window and measured the amplitude of the volume changes at the round window as it bulged out and caved in. He found that at 500 cycles/s (pressure amplitude)/(amplitude of volume displacement) was about 10^9 dyn/cm^5. The volume displacement must have been the same at the oval window as at the round one. The area of the oval window is about $0 \cdot 03$ cm^2 so if the amplitude of volume displacement was x cm^3 the amplitude of vibration of the membrane would average $x/0 \cdot 03$ cm. Hence (pressure amplitude)/(amplitude of vibration at the oval window) was $0 \cdot 03 \times 10^9 = 3 \times 10^7$ dyn/cm^3. Velocity amplitude at a frequency n is $2\pi n$ (amplitude of vibration) and impedance per unit area is (pressure amplitude)/(velocity amplitude). Thus, the unit area impedance of the inner ear at the oval window is $3 \times 10^7/2\pi \times 500 \backsimeq 10^4$ dyn s/cm^3.

The characteristic acoustic impedance of air is only 43 dyn s/cm^3. If sounds arriving at the ear fell directly on the oval window nearly all their energy would be reflected. Very little of the energy would be employed in setting up the vibrations in the inner ear on which hearing depends. In any case, little energy would fall on so small an area.

The sound actually falls on the eardrum, making it vibrate.

The vibrations are transmitted across the air-filled middle ear to the oval window by the ear ossicles. The area of the eardrum is about 20 times the area of the oval window. Energy is collected from a larger area than if sound falls directly on the oval window and since the impedance is spread out over a larger area the impedance per unit area is less. If the difference in area were the only factor that had to be considered the unit area impedance of the ear at the eardrum would be $10^4/20 = 500$ dyn s/cm³.

The ear ossicles are a system of levers with a velocity ratio (see page 13) which is probably about 2. Different investigators have reached different conclusions as to the manner in which the ossicles vibrate and as to the value of the velocity ratio: the value given is from a recent investigation on cats (Guinan and Peake, 1967). If it is correct, the velocity amplitude of the vibrations of the eardrum is twice that of the vibrations of the oval window, but the amplitude of the forces exerted by the eardrum on the ossicles is only half that of the forces exerted by the ossicles on the window. The impedance, (force amplitude)/(velocity amplitude), is only one quarter as large at the eardrum as at the oval window. Allowing for this as well as the difference in area, we can estimate that the unit area impedance of the ear at the eardrum is $500/4 = 125$ dyn s/cm³.

This estimate is based on the false assumption that the impedance of the middle ear is relatively small and can be neglected. In fact, the middle ear has appreciable impedance, due to the mass, stiffness and viscosity of its many components (page 299). The unit area impedance at the eardrum is actually about 1,000 dyn s/cm³ at 500 cycles/s (Zwislocki, 1962). It falls to about 400 dyn s/cm³ at 1,000 cycles/s.

The external ear probably acts like the horn of a phonograph recorder, increasing the area from which sound energy is collected and improving the impedance match with the free air.

An ordinary mammal ear does not work well under water, since it has evolved to match the characteristic acoustic impedance of air. Whales have very peculiar ears, specially adapted for hearing under water (Fraser and Purves, 1960). The modifications involve changes in the proportions of the ear ossicles and in the way they are attached to the eardrum, so that the chain of ossicles has a velocity ratio much less than 1 and the oval window vibrates with a much greater amplitude than the eardrum. In terrestrial mammals the velocity ratio of the ossicles and the disparity in area between drum and window cooperate in making the unit area impedance at the eardrum low. In whales, the low velocity ratio counteracts the disparity in area, and makes the impedance at the eardrum high.

Echo-location by bats

Bats are active by night and most of them feed on small insects such as mosquitoes. They are very adept at catching insects in flight, even in darkness or near darkness. They are apparently just as good at it after they have been blinded. As they fly they emit very high frequency sounds which are inaudible to human ears but which can be detected with suitable equipment. If a bat's mouth is covered so that it cannot make the sounds it collides with objects it would previously have avoided. The same happens if its ears are plugged. It seems that bats find their way about by detecting echoes of the sounds they make, from obstacles around them. There is every indication that they find the insects they eat in the same way (Griffin, 1958, 1960).

What follows applies to the Vespertilionidae, which is the largest family of bats. Some other bats use a different system of echo-location and most of the fruit bats (Megachiroptera) do not use echoes at all.

As a vespertilionid bat such as *Myotis* flies about, it makes a series of very short chirps. Each chirp is 2 or 3 ms long and there is an interval of 70 ms or so between chirps. In each chirp, the pitch falls rapidly down the scale: the frequency is about 110,000 cycles/s (110 kc) at the beginning of the chirp and about 40 kc at the end. People cannot hear sounds above a frequency of about 20 kc.

As a bat approaches an obstacle or chases an insect, it makes shorter chirps, and makes them more rapidly. The chirps may be only 0·3 ms long and 5 ms apart. The frequency is less, around 30 kc, and the pitch falls less during each chirp (Griffin, 1962).

The velocity of sound in air is $3·3 \times 10^4$ cm/s. The first sound wave in a chirp lasting 3 ms would have travelled

$0 \cdot 003 \times 3 \cdot 3 \times 10^4 = 100$ cm by the time the last waves were emitted. If the bat were less than 50 cm from an obstacle the first waves would travel to the obstacle and the echo would return to the bat before the bat had finished the chirp. It might be hard for the bat to hear the first part of the relatively weak echo while it was still producing the intense chirp. The beginning of a chirp lasting $0 \cdot 3$ ms would only travel 10 cm before the end of the chirp. An echo would not be masked by the original chirp unless it came from an obstacle less than 5 cm away.

Cahlander, McCue and Webster (1964) took films of bats catching mealworms which were thrown into the air. They recorded the chirps the bats made at the same time and compared the length of the chirps with the distance of the bat from the mealworm. They obtained two satisfactory films, which they analysed. In each case, the chirps lasted about $1 \cdot 5$ ms when the bat was 50 cm from the mealworm and shortened until they were about $0 \cdot 3$ ms long when the bat was 10 cm from it. Sound travels 50 cm in $1 \cdot 5$ ms and (as we have seen) 10 cm in $0 \cdot 3$ ms. The bats adjusted the lengths of their chirps so that the last waves were emitted at about the time the first ones reached the mealworm. The echo did not get back to the bat until well after the end of the chirp.

If the chirps followed each other in too rapid sequence, a second chirp would be emitted before the echoes of the first had returned. This might complicate the analysis of the data which the bat has to make. When a bat is close to an object the chirps may be only 5 ms apart and there is only time for the sound to travel 170 cm between chirps. Echoes from objects more than 85 cm away would fail to arrive before the next chirp. In cruising flight, well away from obstacles and insects, the chirps may be 70 ms apart. There is time for echoes to return from objects as much as 1,200 cm away, before the next chirp. The rapid sequence of chirps may be best for tracking down a moving insect at close range, but the longer intervals are needed at larger ranges.

The changing rate of chirping as a bat approaches an object has been used to estimate the distance at which objects can be detected. Wires were stretched between the floor and the ceiling of a long room so that a bat flying along the room would have to dodge between them. Bats were released at one end of the room and the rate of chirping was observed as they flew towards the wires. If the diameter of the wires was 1 mm, the rate began to rise when the bat was about 200 cm away. If the diameter was $0 \cdot 2$ mm, the rate began to rise at a distance of about 90 cm. These were presumably the distances at which the wires were detected. Wires down to about $0 \cdot 1$ mm are detected, but finer ones are not: the rate of chirping does not rise and the bats fail to dodge between the wires but collide with them at random.

Bats can plainly tell the direction from which an echo comes. They seem to need both ears for this, for a bat with one ear blocked is nearly as helpless as one with both blocked. They may use either or both of the methods people use for telling the direction of a sound.

A sound arriving from the left is louder on the left side of the head than on the right, because the head is an obstacle for the sound and forms a shadow. The direction an echo comes from could be judged from the difference in its loudness in the two ears. Also, a sound arriving from the left arrives at the left side of the head before the right. The direction an echo comes from could be judged from the difference in the time of its arrival in the two ears. This difference would always be small, for a typical bat's ears are only about 1 cm apart. Sound arriving directly from the left at a velocity of $3 \cdot 3 \times 10^4$ cm/s would reach the left ear $1/3 \cdot 3 \times 10^4$ s $= 0 \cdot 03$ ms before the right one. This is a very short interval, but the high frequency should make it easier to detect. If the frequency was 50 kc the sound arriving at the left ear would be $1 \cdot 5$ cycles out of phase with the sound arriving at the right. The interval would of course be less if the sound was approaching the head obliquely, and zero if it came from directly ahead, or from anywhere else in the median plane.

Both methods of direction finding, by intensity difference and by time difference, are ambiguous. Neither can distinguish, for instance, between different directions in the median plane. This difficulty can be overcome by head movements. If, for instance, an object lies in the median plane when the head is tilted to the left, and in the new median plane when the head is tilted to the right, it must lie on the intersection of the two planes.

Bats can almost certainly estimate the distance of an object, as well as its direction, from the echoes they receive. If they simply flew towards insects with their mouths open it would be enough to know the direction of the insect. If a bat continued flying towards an insect long enough, it would eventually catch it. However, bats do not feed in this way. They usually spread out the membrane between their legs and catch the insect in it, or scoop up the insect in a cupped wing. This involves reaching out at the right moment and must surely depend on knowing how far off the insect is. The distance is probably judged from the interval between making the sound and hearing the echo (Cahlander, McCue and Webster, 1964).

A few species of bats feed on fish. They have large claws on their hind feet and they fly low over water, dipping in their feet from time to time to catch a fish. They feed both in freshwater and in the sea. They can fish in darkness and they make chirps like other bats as they fly. Presumably, they find their prey by echo-location. How do they do it? Nearly all the sound they make must be reflected from the surface of the water. Only $0 \cdot 1\%$ of the sound energy striking the water would enter it, so the intensity of the sound reaching a submerged fish would be very low. The fish's flesh would probably not reflect much sound because it would not differ much from the water in characteristic acoustic impedance, but the swimbladder should reflect sound well. Only $0 \cdot 1\%$ of the reflected sound energy which reached the surface would pass through into the air. The sound travelling from the bat to the fish and back again would have its intensity divided by 1,000 each time it passed through the surface. The intensity of the echo would be only one millionth of the intensity of the echo from a comparable object at the same distance in air. Echo-location of submerged fishes seems scarcely feasible. Experiments by Suthers (1965) seem to show that the bats are, indeed, unable to detect submerged fish, but that they can detect small disturbances of the surface. Fishing bats probably detect ripples on the water surface made by surfacing fish, rather than the fish themselves.

The horseshoe bats (Rhinolophidae) practise echo-location,

but the sounds they use are quite different from the sounds made by Vespertilionidae (Moehres, 1960; Griffin, 1962). The frequencies are similar but the chirps are much longer (from 7 to 100 ms or more) and the intervals between chirps are relatively short. The beginning of an echo must generally return to the bat well before the chirp has ended and the end of an echo must often overlap the next chirp.

Horseshoe bats keep their mouths closed and emit sound through their nostrils. Round the nostrils is the curious "horseshoe" which gives them their name. It is concave and reflects the sound forward in a concentrated beam, like a concave mirror producing a narrow beam of light. It is only a few millimetres across but this is large enough, for the frequency of the sound is 80–100 kc, so the wavelength is only about $0 \cdot 3$–$0 \cdot 4$ cm. Interference between the sound waves from the two nostrils (see page 317) must help to reduce the intensity of sound travelling laterally.

A flying horseshoe bat keeps turning its head in different directions. It directs its beam of sound in various directions in turn, and listens for echoes from each direction. There is no need to make a comparison between the two ears to find the direction of an obstacle, and horseshoe bats can avoid obstacles quite normally after one ear has been blocked. The head may turn through almost 360° during a single chirp, and if information is being collected from all directions the bat can hardly rely on the interval between the beginning of the chirp and the beginning of the echo to judge distance. Perhaps distance is judged from the intensity of the echo.

Interference of sound

Sound waves interfere with each other in the same way as light waves do. Suppose two identical sounds from different sources arrive simultaneously at the same place. If they are in phase, they will add together, so that the pressure amplitude and the amplitude of vibration are each twice as great as if only one of the sounds was happening. The intensity, which is proportional to the square of the amplitude, will be four times as high as for one sound alone. If the sounds are opposite in phase, they will cancel each other out. The pressure amplitude, the amplitude of vibration and the intensity will all be zero.

Fig. 134*a* represents two sources, A and B, which produce identical sounds, in phase with each other. The pressure amplitude is measured at C. If C lies on the broken line the distances AC and BC are equal. The sounds arriving at C will be in phase, and they will add together. If the distance AB is a wavelength or more there are possible positions for C which make AC and BC differ by a whole number of wavelengths, and the sounds again arrive in phase. On the other hand, if AB is half a wavelength or more there are possible positions for C which make AC and BC differ by an odd number of half wavelengths (i.e. by so many and a half wavelengths). The sounds arriving at these positions are opposite in phase and cancel each other out. If AB is exactly half a wavelength the sound intensity at any given distance will be greatest on the broken line and zero along a line at right angles to it, through A and B.

The two nostrils of a horseshoe bat (page 316) are a pair of sound sources set half a wavelength apart. The sounds they emit are identical and in phase because the nostrils are identical conduits carrying sound from a single set of vocal organs. Interference between them will tend to make the sound intensity maximal in the median plane and zero directly lateral to the head. It will help to eliminate stray sound outside the beam directed forwards by the horseshoe.

Fig. 134*b* represents a source D which produces sound which

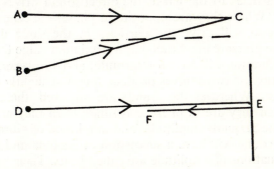

Figure 134. Diagrams illustrating the account of interference of sound waves

is reflected from a wall at E. The pressure amplitude is measured at F. If the distance 2EF is a whole number of wavelengths, the echo will be in phase with the sound arriving direct from D. The pressure amplitude will be the sum of the pressure amplitudes of the direct sound and the echo. If 2EF is an odd number of half wavelengths the echo will be opposite in phase to the direct sound. The pressure amplitude will be the difference of the pressure amplitudes of the direct sound and the echo. The echo can never quite cancel out the direct sound, but it may nearly do so. Along the line DE there are alternating maxima and minima of pressure amplitude. Amplitude fluctuations like this, due to two sets of waves travelling in opposite directions, are known as stationary waves. We have already noted that they can be troublesome in investigations of hearing (page 304).

It has been assumed that the sound is being reflected by a material of higher characteristic impedance than the material it is travelling in. If the reverse were true—if for instance we were dealing with an underwater sound reflected from the water surface—the phase of the sound would be reversed when it was reflected. The direct sound and the echo would have opposite phases when 2EF was a whole number of wavelengths, but they would be in phase when 2EF was an odd number of half wavelengths. Stationary waves would still be formed, but the maxima and minima would be in different places.

The direct sound and the echo are travelling in different directions. Where they are in phase the echo will tend to drive the air particles to the left as the direct sound drives them to the right. Where they are opposite in phase, the echo and the direct sound will drive the particles in the same direction. Where the pressure amplitude has a maximum value the amplitude of vibration will have a minimum, and *vice versa*.

Now suppose two sources produce sound at slightly different frequencies. Wherever the sounds are observed, there will be times when they are in phase and times when they are out of phase. The pressure amplitude (and amplitude of vibration) of the total sound will have a succession of maxima and minima. The fluctuations of amplitude are called beats. Their frequency is the difference between the frequencies of the two sounds.

The frequency of the sound made by a vespertilionid bat falls in the course of a chirp and it has been suggested that these bats might judge distance from the beats between the echo and the lower frequency sound they were emitting when the echo reached them. This hypothesis of course assumes that sound is still being emitted when the echo returns. This does not seem to be the case (page 313).

Pulsating bubbles

When bubbles are blown in water a note can be heard. This is because the bubbles are set pulsating (see Fig. 132*a*) as they are formed. This is a form of free vibration and it depends, like the free vibrations of the hacksaw blade discussed earlier (page 279 onwards), on the interaction of mass and stiffness. The mass is the mass of water that moves radially in and out as the bubble pulsates. The stiffness is due to the pressure of the air changing, as the bubble changes its volume: when the bubble expands, for instance, the pressure in it falls so that there is a pressure difference across the wall of the bubble, tending to compress it again. Meyer (1957) gives a lot of information on pulsating bubbles.

The properties of pulsating bubbles can be described by giving values for mass, stiffness and damping which can be put in the basic equations for free and forced vibration, equations (42) and (45) (pages 281 and 291). Since pressure is force per unit area it is convenient to give mass per unit area. It can be shown that the effective mass per unit area of bubble surface, is approximately equal to $r\rho$, where r is the radius of the bubble and ρ is the density of the surrounding fluid (the approximation only leads to serious error at frequencies well above the resonant frequency). The stiffness per unit area is about $3\gamma P/r$, for bubbles of radius $0\cdot1$ cm and over. γ is the ratio of the specific heat at constant pressure to the specific heat at constant volume, for the gas in the bubble (it is $1\cdot4$ for air or oxygen). P is the pressure. The damping per unit area varies with the size of bubble, but is relatively low. The frequency of free pulsation is therefore about $\sqrt{[S/m]}/2\pi$ or $\sqrt{[3\gamma P/\rho]}/2\pi r$. For an air bubble in shallow water $\gamma = 1\cdot4$, $P = 10^6$ dyn/cm and $\rho = 1$ g/cm^3 so this frequency is $330/r$ cycles/s. A bubble of radius 1 cm will pulsate at about 330 cycles/s and a bubble of radius $0 \cdot 1$ cm at about 3,300 cycles/s.

Now consider a bubble exposed to sound. The fluctuations of pressure alternately compress it and allow it to expand, so that it makes forced pulsations. The pressure amplitude of the sound, p, is the amplitude of the force acting on a unit area of the bubble surface. We already have values for the effective mass and stiffness per unit area. By putting them in equation (46) we obtain an equation giving the amplitude A' of the pulsations

$$A' = p/\sqrt{[[(3\gamma P/r) - 4\pi^2 n^2 \rho r]^2 + (2\pi n K)^2]}$$

The amplitude of vibration A in far field sound of the same pressure amplitude, far from any bubbles, would be $p/2\pi n\rho c$ (equation 51, page 303). This is much less than the amplitude of pulsation, at frequencies within a few octaves of $330/r$ cycles/s. The bubble increases the amplitude by a factor A'/A, and

$$A'/A = 2\pi n r\rho c/\sqrt{[(3\gamma P - 4\pi^2 n^2 r^2 \rho)^2 + (2\pi n r K)^2]} \quad (53)$$

Fig. 135 shows two graphs of A'/A against nr. The upper one is for an air bubble in shallow water, and has been calculated from equation (53). A particular value had to be assumed for the damping, which varies with the size of the bubble. The value chosen is correct for bubbles of radius about $0 \cdot 1$ cm: smaller bubbles would amplify the sound rather less at the resonant frequency, but larger ones would amplify it more.

Swimbladders and hearing

Fish hear by means of otoliths which are affected by the vibrations involved in sound, rather than by the pressure fluctuations. We have just seen how the vibrations in far field sound in water are amplified, around a bubble of gas. Various fish use this effect to increase their sensitivity to sound, by increasing the amplitude of vibration of the otoliths. In most cases the gas involved is the gas in the swimbladder.

The amplitude of vibration around a pulsating bubble is inversely proportional to the square of distance from its centre (see page 300). A swimbladder will increase the amplitude of vibration in response to sound all through the body of a fish,

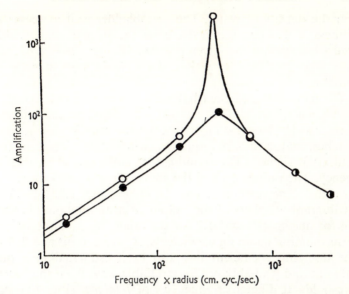

Figure 135. A graph of the amplification of the vibrations of far-field sound against frequency × radius, at the surfaces of a bubble (○) and a swimbladder (●). The amplifications were calculated in the manner described in the text. (From Alexander, 1966b)

but it will have most effect on hearing if it lies close to the ear. Many groups of teleosts including the herring and its relatives (order Clupeiformes) have forward extensions of the swimbladder which contact the ear through openings in the skull. A group of African fish, the Mormyroidei, have vesicles of gas in their ears which develop as extensions of the swimbladder but later get detached from it. The great majority of freshwater fish have a chain of little bones on each side of the body, connecting the swimbladder to the ear. These fish include carps, minnows, catfish etc., and are grouped together as the superorder Ostariophysi. The little bones are called Weberian ossicles.

The amplification at the surface of a swimbladder can be calculated, if equation (53) is modified to apply to the swimbladder instead of a free bubble (Alexander, 1966b). We have to allow for the presence of the swimbladder wall and the body wall. Neither affects the effective mass, for both have densities near the density of water. The swimbladder wall increases the stiffness, if it is taut (i.e. it decreases the volume change which occurs in response to a given pressure change, see Fig. 83, page 199). The swimbladder wall and the body wall increase the damping.

The damping coefficient has been estimated from records of sounds made by two species of fish, which have muscles attached to their swimbladders. They contract the muscles in a series of twitches, and each twitch sets the swimbladder pulsating at its natural frequency. The swimbladder emits sound at this frequency. The pulsations and the sound die gradually away. The logarithmic decrement of the sound pressure, measured from oscillograph displays of the output of a microphone, is about $0 \cdot 4$ for one species and $0 \cdot 7$ for the other. By equation (44) the corresponding damping coefficients, K, are $1 \cdot 6$ mf and $2 \cdot 8$ mf. The damping coefficient has also been estimated in a quite different and probably not very reliable way, from the hearing thresholds at different frequencies of *Ictalurus* (Fig. 133, page 305). The value obtained was 4 mf.

The modified form of equation (53), taking $K = 4$ mf, gave the values for amplification shown by the lower curve in Fig. 135. The main difference between this and the curve for the bubble shown in the same figure, is that the bubble gives much higher amplification at and immediately around the resonant frequency. This is because it is much more lightly damped than the swimbladder. At frequencies well away from the resonant frequency damping has little effect on forced pulsations (see page 291) and the swimbladder gives nearly as much amplification as the bubble. The greater stiffness of the swimbladder gives it a slightly higher resonant frequency than the bubble, and reduces the amplification at low frequencies.

According to the graph, a swimbladder can be expected to amplify the vibrations of sound by a factor of 100 at the resonant frequency (when frequency × radius = 350 cm cycles/s) and by factors of 10 or more when frequency × radius is between about 60 and 2,000 cm cycles/s. A fish with a close connection between the swimbladder and the ear might be expected to hear sounds of $0 \cdot 01$ times the amplitude ($0 \cdot 0001$ times the intensity) which would be just audible to a fish without a swimbladder, if the sound was at the resonant frequency of the swimbladder. In other words, the swimbladder with its connection to the ear might be expected to reduce the threshold of hearing by 40 dB at the resonant frequency. The fish with the connection might be expected to hear sounds of $0 \cdot 1$ times the amplitude ($0 \cdot 01$ times the intensity) just audible to the fish, over a range of 5 octaves. In other words, its threshold over this range might be expected to be 20 dB or more lower. These predictions assume that the amplitude of vibration in the ear is equal to the amplitude of the changes of radius of the pulsating swimbladder, which may not be true of fish with Weberian ossicles (Alexander, 1966b).

Of the fish whose auditory thresholds are shown in Fig. 133, *Holocentrus*, *Carassius* and *Ictalurus* all have connections between the swimbladder and the ear. In *Holocentrus* the anterior end of the swimbladder simply rests against a membranous window in the braincase. *Holocentrus* is sensitive to higher frequencies than *Haemulon* (which has no connection) but it has not got a particularly low threshold at any frequency, and it has a higher threshold than *Equetus* (which has no connection) at all frequencies (Tavolga and Wodinsky, 1963). *Carassius* and *Ictalurus* have the elaborate Weberian apparatus, and extremely low auditory thresholds. Kleerekoper and Roggenkamp (1959) deflated the swimbladder of *Ictalurus* and found that this raised the threshold by small amounts at low frequencies, and by 30 dB at 1,500 cycles/s which was probably close to the resonant frequency of pulsation of the swimbladder (see Alexander, 1966b).

The swimbladder of *Carassius* is separated from the outside of the fish by an ordinary muscular body wall. *Ictalurus* has gaps in the muscle so that the swimbladder lies immediately under the skin. One would expect this to reduce the damping factor and one might expect *Ictalurus* to be, in consequence, more sensitive than *Carassius* to sounds close to the resonant frequency of the swimbladder. No such difference is apparent from Fig. 133.

BIBLIOGRAPHY

ABBOTT, B. C. and WILKIE, D. R. (1953) "The relation between velocity of shortening and the tension-length curve of skeletal muscle", *J. Physiol.* 120: 214–23.

ALEXANDER, R. McN. (1959a) "The physical properties of the swimbladder in intact Cypriniformes", *J. exp. Biol.* 36: 315–32.

ALEXANDER, R. McN. (1959b) "The densities of Cyprinidae", *J. exp. Biol.* 36: 333–40.

ALEXANDER, R. McN. (1959c) "The physical properties of the swim-bladders of fish other than Cypriniformes", *J. exp. Biol.* 36: 347–55.

ALEXANDER, R. McN. (1961) "The physical properties of the swim-bladders of some South American Cypriniformes", *J. exp. Biol.* 38: 403–10.

ALEXANDER, R. McN. (1962) "Visco-elastic properties of the body wall of sea anemones", *J. exp. Biol.* 39: 373–86.

ALEXANDER, R. McN. (1964a) "Visco-elastic properties of the mesogloea of jellyfish", *J. exp. Biol.* 41: 363–9.

ALEXANDER, R. McN. (1964b) "Adaptation in the skulls and cranial muscles of South American characinoid fish", *J. Linn. Soc. (Zool.)* 45: 169–90.

ALEXANDER, R. McN. (1965) "The lift produced by the heterocercal tails of Selachii", *J. exp. Biol.* 43: 131–8.

ALEXANDER, R. McN. (1966a) "Structure and function in the catfish", *J. Zool. Lond.* 148: 88–152.

ALEXANDER, R. McN. (1966b) "Physical aspects of swimbladder function", *Biol. Rev.* 41: 141–76.

ALEXANDER, R. McN. (1966c) "Rubber-like properties of the inner hinge-ligament of Pectinidae", *J. exp. Biol.* 44: 119–30.

ALEXANDER, R. McN. (1967a) "Mechanisms of the jaws of some atherini-form fish", *J. Zool. Lond.* 151: 233–55.

ALEXANDER, R. McN. (1967b) *Functional design in fishes* (Hutchinson, London).

ALLEN, G., BIANCHI, U. and PRICE, C. (1963) "Thermodynamics of elasticity of natural rubber", *Trans. Faraday Soc.* 59: 2,493–502.

ANDERSEN, S. O. (1966) "Covalent cross-links in a structural protein, resilin", *Acta physiol. Scand.* 66 (suppl. 263): 1–81.

ANDERSEN, S. O. (1967) "Isolation of a new type of cross link from the hinge ligament protein of molluscs", *Nature, Lond.* 216: 1,029–30.

ANDERSEN, S. O. and WEIS-FOGH, T. (1964) "Resilin. A rubberlike protein in insect cuticle", *Adv. Insect Physiol.* 2: 1–65.

ARDRAN, G. M., KEMP, F. H. and RIDE, W. D. L. (1958) "A radiographic analysis of mastication and swallowing in the domestic rabbit, *Oryctolagus cuniculus* (L.)", *Proc. zool. Soc., Lond.* 130: 257–74.

ASCENZI, A., BONUCCI, E. and CHECCUCCI, A. (1966) "The tensile properties of single osteons studied using a microwave extensimeter", in Evans, F. G. (ed.) *Studies on the anatomy and function of bone and joints* (Springer, Berlin).

BADOUX, D. M. (1965) "Some notes on the functional anatomy of *Macropus giganteus* Zimm with general remarks on the mechanics of bipedal leaping", *Acta Anat.* 62: 418–33.

BAINBRIDGE, R. (1961) "Problems of fish locomotion", *Symp. zool. Soc., Lond.* 5: 13–32.

BARNETT, C. H. (1954) "The structure and functions of fibrocartilages within vertebrate joints", *J. Anat.* 88: 363–8.

BARNETT, C. H. and COBBOLD, A. F. (1962) "Lubrication within living joints", *J. Bone Jt Surg.* 44 B: 662–74.

BARNETT, C. H., DAVIES, D. V. and MACCONNAILL, M. A. (1961) *Synovial joints, their structure and mechanics* (Longmans, London).

BARNETT, C. H. and LEWIS, O. J. (1958) "The evolution of some traction epiphyses in birds and mammals", *J. Anat.* 92: 593–601.

BASKIN, R. J. and PAOLINI, P. J. (1966) "Muscle volume changes", *J. gen. Physiol.* 49: 387–404.

BASMAJIAN, J. V. (1962) *Muscles alive: their functions revealed by electromyography* (Williams and Wilkins, Baltimore).

BATHAM, E. J. and PANTIN, C. F. A. (1950a) "Muscular and hydrostatic action in the sea-anemone *Metridium senile* (L.)", *J. exp. Biol.* 27: 264–89.

BATHAM, E. J. and PANTIN, C. F. A. (1950b) "Phases of activity in the sea anemone *Metridium senile* (L.) and their relation to external stimuli", *J. exp. Biol.* 27: 377–99.

BECHT, G. (1953) "Comparative biologic-anatomical researches on mastication in some mammals", *Proc. K. Med. Acad. Wet.* 56: 508–27.

BEGGS, J. S. (1955) *Mechanism.* (McGraw-Hill, New York).

BÉKÉSY, G. von (1949) "The vibration of the cochlear partition in anatomical preparations and in models of the inner ear", *J. acoust. Soc. Am.* 21: 233–45.

BELL, G. H., DAVIDSON, J. N. and SCARBOROUGH, H. (1965) *Textbook of physiology and biochemistry* ed. 6 (Livingstone, Edinburgh).

BENNET-CLARK, H. C. and LUCEY, E. C. A. (1967) "The jump of the flea: a study of the energetics and a model of the mechanism", *J. exp. Biol.* 47: 59–76.

BENNINGHOFF, A. and ROLLHÄUSER, H. (1952) "Zur inneren Mechanik des gefiederten Muskels", *Pflügers Arch. ges. Physiol.* 254: 527–48.

BERGEL, D. H. (1961) "The static elastic properties of the arterial wall" *J. Physiol.* 156: 445–57.

BEVAN, T. (1956) *The theory of machines. A text-book for engineering students* ed. 3 (Longmans, London).

BOCK, W. J. (1964) "Kinetics of the avian skull", *J. Morph.* 114: 1–42.

BOETTIGER, E. G. and FURSHPAN, E. (1952) "The mechanics of flight movements in Diptera", *Biol. Bull., Woods Hole* 102: 200–11.

BOTTEMA, O. (1950) "On Grübler's formulae for mechanisms", *Appl. Sci. Res.* A2: 162–4.

BOURNE, G. H. (ed.) (1956) *The biochemistry and physiology of bone* (Academic Press, New York).

BRAZIER, L. G. (1927) "On the flexure of thin cylindrical shells and other 'thin' sections", *Proc. R. Soc.* A, 116: 104–14.

BROWN, R. H. J. (1948) "The flight of birds. The flapping cycle of the pigeon", *J. exp. Biol.* 25: 322–33.

BROWN, R. H. J. (1953) "The flight of birds. II. Wing function in relation to flight speed", *J. exp. Biol.* 30: 90–103.

BROWN, R. H. J. (1963a) "Jumping arthropods", *Times Sci. Rev.*, summer 1963: 6–7.

BROWN, R. H. J. (1963b) "The flight of birds", *Biol. Rev.* 38: 460–89.

BROWN, R. H. J. (1967) "Mechanism of locust jumping", *Nature, Lond.* 214: 939.

BRUNET, P. C. J. (1967) "Sclerotins", *Endeavour* 26: 68–74.

BUCHTHAL, F. and KAISER, E. (1951) "The rheology of the cross striated muscle fibre with particular reference to isotonic conditions", *Dan. Biol. Medd.* 21(7): 1–318.

BUECHE, F. (1958) "Tensile strength of filled GR-S vulcanizates", *J. Polymer Sci.* 33: 259–71.

BULLEID, C. H. (1922) "Kinematics of machinery", *in* Glazebrook, R. (edit.) *A dictionary of applied physics* 1: 542–50.

CAHLANDER, D. A., MCCUE, J. J. G. and WEBSTER, F. A. (1964) "The determination of distance by echolocating bats", *Nature, Lond.* 201: 544–6.

CAPLAN, S. R. (1966) "A characteristic of self-regulated linear energy converters. The Hill force-velocity relation for muscle", *J. theoret. Biol.* 11: 63–86.

CHAPMAN, G. (1950) "Of the movement of worms", *J. exp. Biol.* 27: 29–39.

CHAPMAN, G. (1958) "The hydrostatic skeleton in the invertebrates", *Biol. Rev.* 33: 338–71.

CHAPMAN, G. (1966) "The structure and functions of the mesogloea", *Symp. zool. Soc., Lond.* 16: 147–68.

CHAPMAN, G. and NEWELL, G. E. (1947) "The rôle of the body fluid in relation to movement in soft-bodied invertebrates. I. The burrowing of *Arenicola*", *Proc. R. Soc.* B. 134: 431–55.

CHARNLEY, J. (1959) "The lubrication of animal joints", *New Scientist* 6: 60–1.

CIFERRI, A. (1963) "The $\alpha \leftrightarrows \beta$ transformation in keratin", *Trans. Faraday Soc.* 59: 562–9.

CLARK, R. B. (1964) *Dynamics in metazoan evolution. The origin of the coelom and segments* (Oxford University Press).

CLARK, R. B. and COWEY, J. B. (1958) "Factors controlling the change of shape of certain nemertean and turbellarian worms", *J. exp. Biol.* 35: 731–48.

CLEMENTS, J. A. (1962) "Surface tension in the lungs", *Scient. Am.* 207(6): 121–30.

CLEMENTS, J. A., BROWN, E. S. and JOHNSON, R. P. (1958) "Pulmonary surface tension and the mucous lining of the lungs: some theoretical considerations', *J. appl. Physiol.* 12: 262–6.

CLEMENTS, J. A., HUSTEAD, R. F., JOHNSON, R. P. and GRIBETZ, I. (1961) "Pulmonary surface tension and alveolar stability", *J. appl. Physiol.* 16: 444–50.

COTTRELL, A. H. (1964) *The mechanical properties of matter* (Wiley, New York).

COWAN, P. M., NORTH, A. C. T. and RANDALL, J. T. (1955) "X-ray diffraction studies of collagen fibres", *Symp. Soc. exp. Biol.* 9: 115–26.

COWEY, J. B. (1952) "The structure and function of the basement membrane muscle system in *Amphiporus lactifloreus* (Nemertea)", *Quart. Jl microsc. Sci.* 93: 1–15.

CRISP, D. J. (1950) "The stability of structures at a fluid interface", *Trans. Faraday Soc.* 46: 228–35.

CRISP, D. J. (1964) "Plastron respiration", *Rec. Prog. Surface Sci.* 2: 377–425.

CROMPTON, A. W. (1963) "On the lower jaw of *Diarthrognathus* and the origin of the mammalian lower jaw", *Proc. zool. Soc. Lond.* 140: 697–753.

CURREY, J. D. (1959) "Differences in the tensile strength of bone of different histological types", *J. Anat.* 93: 87–95.

CURREY, J. D. (1962a) "Stress concentrations in bone", *Quart. Jl microsc. Sci.* 103: 111–33.

CURREY, J. D. (1962b) "Strength of bone", *Nature, Lond.* 195: 513–4.

CURREY, J. D. (1964) "Three analogies to explain the mechanical properties of bone", *Biorheology* 2: 1–10.

CURREY, J. D. (1967) "The failure of exoskeletons and endoskeletons", *J. Morph.* 123: 1–16.

CURREY, J. D. (1968) "The effect of protection on the impact strength of rabbit's bones", *Acta Anat.* (in the press).

CURREY, J. D. and NICHOLS, D. (1967) "Absence of organic phase in echinoderm calcite", *Nature, Lond.* 214: 81–3.

DAVIES, D. V. (1966) "Synovial fluid as a lubricant", *Fed. Proc.* 25: 1,069–76.

DAVIES, J. T. and RIDEAL, E. K. (1961) *Interfacial phenomena* (Academic Press, New York).

DAVIES, R. E. (1963) "A molecular theory of muscle contraction: calcium-dependent contractions with hydrogen bond formation plus ATP-dependent extensions of part of the myosin-actin cross-bridge", *Nature, Lond.* 199: 1,068–74.

DENTON, E. J. and GILPIN-BROWN, J. B. (1961a) "The buoyancy of the cuttlefish, *Sepia officinalis* (L.)", *J. mar. biol. Assoc. U.K.* 41: 319–42.

DENTON, E. J. and GILPIN-BROWN, J. B. (1961b) "The distribution of gas and liquid within the cuttlebone", *J. mar. biol. Assoc. U.K.* 41: 365–81.

DENTON, E. J. and GILPIN-BROWN, J. B. (1966) "On the buoyancy of the pearly nautilus", *J. mar. biol. Assoc. U.K.* 46: 723–59.

DENTON, E. J., GILPIN-BROWN, J. B. and HOWARTH, J. V. (1961) "The osmotic mechanism of the cuttlebone", *J. mar. biol. Assoc. U.K.* 41: 351–63.

DICKINSON, S. (1929) "The efficiency of bicycle-pedalling, as affected by speed and load", *J. Physiol.* 67: 242–55.

DINTENFASS, L. (1963) "Lubrication in joints: a theoretical analysis", *J. Bone It. Surg.* 45A: 1,241–56.

DONALDSON, P. E. K. and others (1958) *Electronic apparatus for biological research* (Butterworth, London).

DYSON, G. H. G. (1962) *The mechanics of athletics* (University of London Press).

EDMAN, K. A. P. (1966) "The relation between sarcomere length and active tension in isolated semitendinosus fibres of the frog", *J. Physiol.* 183: 407–17.

ELLIOTT, D. H. (1965) "Structure and function of mammalian tendon", *Biol. Rev.* 40: 392–421.

ENDO, B. (1965) "Distribution of stress and strain produced in the human facial skeleton by the masticatory force", *J. anthrop. Soc. Nippon* 73: 123–36.

EVANS, F. G. (1957) *Stress and strain in bones* (Thomas, Springfield Illinois).

FERRY, J. D. (1961) *Viscoelastic properties of polymers* (Wiley, New York).

FEUGHELMAN, M. (1963) "Free-energy difference between the alpha and beta states in keratin", *Nature, Lond.* 200: 127–9.

FRASER, F. C. and PURVES, P. E. (1960) "Anatomy and function of the cetacean ear", *Proc. R. Soc.* B 152: 62–77.

FRAZZETTA, T. H. (1962) "A functional consideration of cranial kinesis in lizards", *J. Morph.* 111: 287–319.

FRAZZETTA, T. H. (1966) "Studies on the morphology and function of the skull in the Boidae (Serpentes). Part II. Morphology and function of the jaw apparatus in *Python sebae* and *Python molurus*", *J. Morph.* 118: 217–96.

GADD, G. P. (1963) "Some hydrodynamical aspects of swimming", *Nat. Phys. Lab., Ship. Div. Ship. Rept.* 45: 1–22.

GADD, G. E. (1966) "Reduction of turbulent friction in liquids by dissolved additives", *Nature, Lond.* 212: 874–7.

GANS, C. and BOCK, W. J. (1965) "The functional significance of muscle architecture", *Ergebn. Anat. EntwGesch.* 38: 115–42.

GLASSTONE, S. and LEWIS, D. (1962) *Elements of physical chemistry* ed. 2 (Macmillan, London).

GORDON, A. M., HUXLEY, A. F. and JULIAN, F. J. (1966) "The variation in isometric tension with sarcomere length in vertebrate muscle fibres", *J. Physiol.* 184: 170–92.

GRAY, J. (1933) "Studies in animal locomotion. I. The movement of fish with special reference to the eel", *J. exp. Biol.* 10: 88–104.

GRAY, J. (1936) "Studies in animal locomotion. VI. The propulsive powers of the dolphin", *J. exp. Biol.* 13: 192–9.

GRAY, J. (1953) *How animals move* (Cambridge University Press).

GRAY, J. and LISSMANN, H. W. (1938a) "Studies in animal locomotion. VII. Locomotory reflexes in the earthworm", *J. exp. Biol.* 15: 506–16.

GRAY, J. and LISSMANN, H. W. (1938b) "An apparatus for measuring the propulsive forces of the locomotory muscles of the earthworm and other animals", *J. exp. Biol.* 15: 518–21.

GREENEWALT, C. H. (1962) "Dimensional relationships for flying animals", *Smithson. misc. Collns* 144(2): 1–46.

GRIFFIN, D. R. (1958) *Listening in the dark. The acoustic orientation of bats and men* (Yale University Press, New Haven).

GRIFFIN, D. R. (1960) *Echoes of bats and men* (Heinemann, London).

GRIFFIN, D. R. (1962) "Comparative studies of the orientation sounds of bats", *Symp. zool. Soc., Lond.* 7: 61–72.

GUEST, M. M., BOND, T. P., COOPER, R. G. and DERRICK, J. R. (1963) "Red blood cells: change in shape in capillaries", *Science* 142: 1,319–21.

GUINAN, J. J. and PEAKE, W. T. (1967) "Middle-ear characteristics of anaesthetized cats", *J. acoust. Soc. Am.* 41: 1,237–61.

HALE, L. J. (1965) *Biological Laboratory Data* (Methuen, London).

HALL-CRAGGS, E. C. B. (1965) "An analysis of the jump of the Lesser Galago (*Galago senegalensis*)", *J. Zool., Lond.* 147: 20–9.

HAMMOND, R. A. (1966) "Changes of internal hydrostatic pressure and body shape in *Acanthocephalus ranae*", *J. exp. Biol.* 45: 197–202.

HARKNESS, M. L. R. and HARKNESS, R. D. (1959a) "Changes in the physical properties of the uterine cervix of the rat during pregnancy", *J. Physiol.* 148: 524–47.

HARKNESS, M. L. R. and HARKNESS, R. D. (1959b) "Effect of enzymes on mechanical properties of tissues", *Nature, Lond.* 183: 1,821–2.

HARKNESS, M. L. R., HARKNESS, R. D. and McDONALD, D. A. (1957) "The collagen and elastin content of the arterial wall in the dog", *Proc. R. Soc.* B 146: 541–51.

HARKNESS, R. D. (1961) "Biological functions of collagen", *Biol. Rev.* 36: 399–463.

HARKNESS, R. D. (1966) "Collagen", *Science Progr.* 54: 257–74.

HARRIS, G. G. (1964) "Considerations on the physics of sound production by fishes", *in* Tavolga, W. N. (ed.) *Marine Bio-Acoustics* 233–47 (Pergamon, Oxford).

HARRIS, J. E. (1936) "The role of the fins in the equilibrium of the swimming fish. I. Wind-tunnel tests on a model of *Mustelus canis* (Mitchell)" *J. exp. Biol.* 13: 476–93.

HEARLE, J. W. S. (1958) "A fringed fibril theory of structure in crystalline polymers", *J. Polymer Sci.* 28: 432–5.

HEARLE, J. W. S. (1963a) "The fine structure of fibers and crystalline polymers. I. Fringed fibril structure", *J. appl. Polymer Sci.* 7: 1,175–92.

HEARLE, J. W. S. (1963b) "The fine structure of fibers and crystalline polymers. III. Interpretation of the mechanical properties of fibers", *J. appl. Polymer Sci.* 7: 1,207–23.

HEARLE, J. W. S. and PETERS, R. H. (edit.) (1963) *Fibre structure* (The Textile Institute and Butterworth, Manchester and London).

HERTEL, H. (1966) *Structure, form, movement* (Reinhold, New York).

HILL, A. V. (1938) "The heat of shortening and dynamic constants of muscle", *Proc. R. Soc.* B 126: 136–95.

HILL, A. V. (1950) "The dimensions of animals and their muscular dynamics", *Science Progr.* 38: 209–30.

HILL, A. V. (1956) "Thermodynamics of muscle", *Brit. med. Bull.* 12: 174–6.

HODGMAN, C. D. (ed.) (1965) *Handbook of Chemistry and Physics* ed. 46 (Chemical Rubber Co., Cleveland, Ohio).

HOEVE, C. A. J. and FLORY, P. J. (1958) "The elastic properties of elastin", *J. Am. chem. Soc.* 80: 6,523–6.

HOEVE, C. A. J. and FLORY, P. J. (1962) "Elasticity of crosslinked amorphous polymers in swelling equilibrium with diluents", *J. Polymer Sci.* 60: 155–64.

HOLDGATE, M. W. (1955) "The wetting of insect cuticles by water", *J. exp. Biol.* 32: 591–617.

HOLLIDAY, L. (ed.) (1966) *Composite materials* (Elsevier, Amsterdam).

HOLWILL, M. E. J. (1966) "Physical aspects of flagellar movement", *Physiol. Rev.* 46: 696–785.

HOYLE, G. (1955) "Neuromuscular mechanisms of a locust skeletal muscle", *Proc. R. Soc.* B 143: 343–67.

HUGHES, G. M. (1958) "The co-ordination of insect movements. III. Swimming in *Dytiscus*, *Hydrophilus*, and a dragonfly nymph", *J. exp. Biol.* 35: 567–83.

HUGHES, G. M. and SHELTON, G. (1962) "Respiratory mechanisms and their nervous control in fish", *Adv. comp. Physiol. Biochem.* 1: 275–364.

IIZUKA, E. (1966) "Mechanism of fiber formation by the silkworm, *Bombyx mori* L.", *Biorheology* 3: 141–52.

INGELS, N. P. and THOMPSON, N. P. (1966) "An electrokinematic theory of muscle contraction', *Nature, Lond.* 211: 1,032–5.

JACOBS, D. W. and TAVOLGA, W. N. (1967) "Acoustic intensity limens in the goldfish", *Anim. Behav.* 15: 324–35.

JAMESON, W. (1958) *The wandering albatross* (Hart-Davis, London).

JENSEN, M. (1956) "Biology and physics of locust flight. III. The aerodynamics of locust flight", *Phil. Trans.* B 239: 511–52.

JENSEN, M. and WEIS-FOGH, T. (1962) "Biology and physics of locust flight. V. Strength and elasticity of locust cuticle", *Phil.Trans.* B 245: 137–69.

JEWELL, B. R. and WILKIE, D. R. (1958) "An analysis of the mechanical components in frogs' striated muscle", *J. Physiol.* 143: 515–40.

JONES, F. R. H. (1951) "The swimbladder and the vertical movement of teleostean fishes. I. Physical factors", *J. exp. Biol.* 28: 553–66.

JONES, F. R. H. (1952) "The swimbladder and the vertical movements of teleostean fishes. II. The restriction to rapid and slow movements", *J. exp. Biol.* 29: 94–109.

KELLY, R. E. and RICE, R. V. (1967) "Abductin: a rubber-like protein from the internal triangular hinge ligament of *Pecten*", *Science* 155: 208–10.

KLEEREKOPER, H. and ROGGENKAMP, P. A. (1959) "An experimental study on the effect of the swimbladder on the hearing sensitivity in *Ameiurus nebulosus nebulosus* (Lesueur)", *Can. J. Zool.* 37: 1–8.

KRAMER, M. O. (1960) "The dolphin's secret", *New Scientist* 7: 1,118–20.

KRAMER, M. O. (1965) "Hydrodynamics of the dolphin", *Adv. Hydrosci* 2: 111–30.

LANDOLT, H. and BÖRNSTEIN, R. (1955) *Zahlenwerte und Funktionenaus. Physik, Chemie, Astronomie, Geophysik und Technik*, ed. 6, vol. 4, part 1 (Springer, Berlin).

LANG, T. G. (1966) "Hydrodynamic analysis of dolphin fin profiles", *Nature, Lond.* 209: 1,110–1.

LANG, T. G. and NORRIS, K. S. (1966) "Swimming speed of a Pacific bottlenose porpoise", *Science* 151: 588–90.

LANG, T. G. and PRYOR, K. (1966) "Hydrodynamic performance of porpoises (*Stenella attenuata*)' *Science* 152: 531–3.

LASIEWSKI, R. C. (1963) "Oxygen consumption of torpid, resting, active and flying hummingbirds", *Physiol. Zool.* 36: 122–40.

LEWIS, P. R. and McCUTCHEN, C. W. (1959) "Experimental evidence for weeping lubrication in mammalian joints", *Nature, Lond.* 184: 1,285.

LIGHTHILL, M. J. (1960) "Note on the swimming of slender fish", *J. Fluid Mech.* 9: 305–17.

LITTLER, T. S. (1965) *The physics of the ear* (Pergamon, Oxford).

LODGE, A. S. (1964) *Elastic liquids* (Academic Press, London).

LOWENSTAM, H. A. (1962a) "Magnetite in denticle capping in recent chitons (Polyplacophora)", *Bull. geol. Soc. Am.* 73: 435–8.

LOWENSTAM, H. A. (1962b) "Goethite in radular teeth of recent marine gastropods", *Science* 137: 279–80.

LOWNDES, A. G. (1942) "The displacement method of weighing living aquatic organisms", *J. mar. biol. Ass. U.K.* 25: 555–74.

LOWNDES, A. G. (1955) "Density of fishes. Some notes on the swimming of fish to be correlated with density, sinking factor and load carried", *Ann. Mag. nat. Hist.* (12) 8: 241–56.

McCUTCHEN, C. W. (1959) "Sponge-hydrostatic and weeping bearings", *Nature, Lond.* 184: 1,284–5.

McCUTCHEN, C. W. (1962a) "The frictional properties of animal joints", *Wear* 5: 1–17.

McCutchen, C. W. (1962b) "Animal joints and weeping lubrication", *New Scientist* 15: 412–15.

McCutchen, C. W. (1966) "Boundary lubrication by synovial fluid: demonstration and possible osmatic explanation", *Fed. Proc.* 25: 1,061–8.

Machin, K. E. and Pringle, J. W. S. (1959) "The physiology of insect fibrillar muscle. II. Mechanical properties of a beetle flight muscle", *Proc. R. Soc.* B 151: 204–25.

Manton, S. M. (1965) "The evolution of arthropodan locomotory mechanisms, part 8. Functional requirements and body design in Chilopoda", *J. Linn. Soc. (Zool.)* 46: 251–484.

Marino, A. A. and Becker, R. O. (1967) "Evidence for direct physical bonding between the collagen fibres and apatite crystals in bone", *Nature, Lond.* 213: 697–8.

Märkel, K. (1964) "Modell-Untersuchungen zur Klärung der Arbeitsweise der Gastropodenradula", *Verh. dt. Zool. Ges.* 1964: 232–43.

Meyer, E. (1957) "Air bubbles in water", *in Technical aspects of sound* 2: 222–39, ed. E. G. Richardson (Elsevier, Amsterdam).

Moehres, F. P. (1960) "Sonic orientation of bats and other animals", *Symp. zool. Soc., Lond.* 3: 57–66.

Nachtigall, W. (1960) "Über Kinematik, Dynamik und Energetik des Schwimmens einheimischer Dytisciden", *Z. vergl. Physiol.* 43: 48–118.

Nachtigall, W. (1966) "Die Kinematik der Schlagflügelbewegungen von Dipteren", *Z. vergl. Physiol.* 52: 155–211.

Nachtigall, W. (1967) "Aerodynamische Messungen am Tragflügelsystem segelner Schmetterlinge", *Z. vergl. Physiol.* 54: 210–31.

Nachtigall, W. and Bilo, D. (1965) "Die Strömungsmechanik des *Dytiscus*-Rumpfes", *Z. vergl. Physiol.* 50: 371–401.

Nachtigall, W. and Wieser, J. (1966) "Profilmessungen am Taubenflügel", *Z. vergl. Physiol.* 52: 333–46.

National Research Council of the U.S.A. (1928) *International critical tables of numerical data, physics, chemistry and technology* 3 (McGraw-Hill, New York).

Nelkon, M. and Parker, P. (1965). *Advanced level physics* ed. 2 (Heinemann, London).

Newell, G. E. (1950) "The role of the coelomic fluid in the movements of earthworms", *J. exp. Biol.* 27: 110–21.

Newman, B. G. (1958) "Soaring and gliding flight in the black vulture", *J. exp. Biol.* 35: 280–5.

Ogston, A. G. and Stanier, J. E. (1953a) "The physiological functions of hyaluronic acid in synovial fluid; viscous, elastic and lubricant properties", *J. Physiol.* 119: 244–52.

Ogston, A. G. and Stanier, J. E. (1953b) "Some effects of hyaluronidase on the hyaluronic acid content of ox synovial fluid, and their bearing on the investigation of pathological fluids", *J. Physiol.* 119: 253–8.

Oth, J. F. M., Dumitru, E. T., Spurr, O. K. and Flory, P. J. (1957) "Phase equilibrium in the hydrothermal shrinkage of collagen", *J. Am. chem. Soc.* 79: 3,288–9.

PARKE, S. (1966) "Logarithmic decrement at high damping", *Brit. J. appl. Phys.* 17: 271–3.

PARRY, D. A. (1949) "The swimming of whales and a discussion of Gray's paradox", *J. exp. Biol.* 26: 24–34.

PARRY, D. A. (1965) "The signal generated by an insect in a spider's web", *J. exp. Biol.* 43: 185–92.

PARRY, D. A. and BROWN, R. H. J. (1959a) "The hydraulic mechanism of the spider leg", *J. exp. Biol.* 36: 423–33.

PARRY, D. A. and Brown, R. H. J. (1959b) "The jumping mechanism of salticid spiders", *J. exp. Biol.* 36: 654–64.

PATTLE, R. E. (1965) "Surface lining of lung alveoli", *Physiol. Rev.* 45: 48–79.

PAUWELS, F. (1948) "Die Bedeutung der Bauprinzipien für die Beanspruchung der Röhrenknochen", *Z. Anat. EntwGes.* 114: 129–66.

PEARSON, O. P. (1950) "The metabolism of hummingbirds", *Condor* 52: 145–52.

PENNYCUICK, C. J. (1960) "Gliding flight of the fulmar petrel", *J. exp. Biol.* 37: 330–8.

PENNYCUICK, C. J. (1967) "The strength of the pigeon's wing bones in relation to their function", *J. exp. Biol.* 46: 219–33.

PENNYCUICK, C. J. and PARKER, G. A. (1966) "Structural limitations on the power output of the pigeon's flight muscles", *J. exp. Biol.* 45: 489–98.

PENNYCUICK, C. J. and WEBBE, D. (1959) "Observations on the fulmar in Spitsbergen, *British Birds* 52: 321–32.

PETERSON, R. E. (1953) *Stress concentration design factors* (Wiley, New York).

PINNOCK, P. R. and WARD, I. M. (1966) "Mechanical and optical anisotropy in polypropylene fibres", *Brit. J. appl. Phys.* 17: 575–86.

PRANDTL, L. (1952) *Essentials of fluid dynamics* (Blackie, London).

PRINGLE, J. W. S. (1957) *Insect flight* (Cambridge University Press).

PRINGLE, J. W. S. (1967) "The contractile mechanism of insect fibrillar muscle", *Progr. Biophys. mol. Biol.* 17: 1–60.

PROTHERO, J. and BURTON, A. C. (1961) "The physics of blood flow in capillaries. I. The nature of the motion", *Biophys. J.* 1: 565–80.

PRYOR, M. G. M. (1962) 'Sclerotization", *in* Florkin, M. and Mason, H. S. (edd.) *Comparative Biochemistry* 4: 371–96 (Academic Press, New York).

PURVES, P. E. (1963) "Locomotion in whales", *Nature, Lond.* 197: 334–7.

RACK, P. M. H. (1966) "The behaviour of a mammalian muscle during sinusoidal stretching", *J. Physiol.* 183: 1–14.

RADFORD, E. P. (1957) "Recent studies of mechanical properties of mammalian lungs", *in* Remington, J. W. (edit.) *Tissue Elasticity* (American Physiological Society, Washington).

RAMSEY, A. S. (1941) *Statics. A text-book for the use of the higher divisions in schools and for first year students at the universities* ed. 2 (Cambridge University Press).

RAMSEY, A. S. (1946) *Hydrostatics. A text-book for the use of first year students at the universities and for the higher divisions in schools* ed. 2 (Cambridge University Press).

RANDALL, R. H. (1951) *An introduction to acoustics* (Addison-Wesley, Reading, Mass.).

RASPET, A. (1960) "Biophysics of bird flight", *Ann. Rept. Smithson Inst.* 1960, 405–25.

RITCHIE, P. D. (edit.) (1965) *Physics of plastics* (Iliffe, London).

ROMER, A. S. (1956) *Osteology of the reptiles* (University of Chicago Press).

RYDELL, N. (1966) "Intravital measurements of forces acting on the hip joint", *in* Evans, F. G. (ed.) *Studies on the anatomy and function of bone and joints* (Springer, Berlin).

SAVORY, T. H. (1952) *The spider's web* (Warne, London).

SCHMITZ, F. W. (1960) *Aerodynamik des Flügmodells* ed. 4 (Lange, Duisburg).

SERAFINI-FRACASSINI, A. and TRISTRAM, G. R. (1966) "Electron microscopic study and amino acid analysis on human aortic elastin", *Proc. R. Soc. Edin.* 69: 334–44.

SMITH, J. M. and SAVAGE, R. J. G. (1956) "Some locomotory adaptations in mammals", *J. Linn. Soc. (Zool.)* 42: 603–22.

SMITH, J. M. and SAVAGE, R. J. G. (1959) "The mechanics of mammalian jaws", *School Science Rev.* 40: 289–301.

SMITH, J. W. (1960a) "Collagen fibre patterns in mammalian bone", *J. Anat.* 94: 329–44.

SMITH, J. W. (1960b) "The arrangement of collagen fibres in human secondary osteones", *J. Bone Jt Surg.* 42 B: 588–605.

SMITH, J. W. (1962a) "The relationship of epiphysial plates to stress in some bones of the lower limb", *J. Anat.* 96: 58–78.

SMITH, J. W. (1962b) "The structure and stress relations of fibrous epiphysial plates", *J. Anat.* 96: 209–25.

SMITH, J. W. and WALMSLEY, R. (1959) "Factors affecting the elasticity of bone", *J. Anat.* 93: 503–23.

SNODGRASS, R. E. (1935) *Principles of insect morphology* (McGraw-Hill, New York).

SOTAVALTA, O. (1952) "The essential factor in regulating the wing-stroke frequency of insects in wing mutilation and loading experiments and in experiments at subatmospheric pressure", *Ann. zool. Soc. "Vanamo"* 15(2): 1–67.

STEVEN, G. A. (1950) "Swimming of dolphins", *Science Progr.* 38: 524–5.

SUTHERS, R. A. (1965) "Acoustic orientation by fish-catching bats", *J. exp. Zool.* 158: 319–48.

TAVOLGA, W. N. and WODINSKY, J. (1963) "Auditory capacities in fish. Pure tone thresholds in nine species of marine teleosts", *Bull. Am. Mus. nat. Hist.* 126: 177–240.

TAYLOR, G. (1952) "Analysis of the swimming of long and narrow animals", *Proc. R. Soc.* A 214: 158–83.

THOM, A. and SWART, P. (1940) "The forces on an aerofoil at very low speeds", *J. R. aero. Soc.* 44: 761–70.

THORPE, W. H. and CRISP, D. J. (1947) "Studies on plastron respiration. I. The biology of *Aphelocheirus* [Hemiptera, Aphelocheiridae (Naucoridae)] and the mechanism of plastron retention", *J. exp. Biol.* 24: 227–69.

THORPE, W. H. and CRISP, D. J. (1949) "Studies on plastron respiration. IV. Plastron respiration in the Coleoptera", *J. exp. Biol.* 26: 219–60.

TIETJENS, O. G. (1957) *Applied hydro- and aerodynamics* (Dover, New York).

TIMOSHENKO, S. (1936) *Theory of elastic stability* (McGraw-Hill, New York).

TRELOAR, L. R. G. (1958) *The physics of rubber elasticity* ed. 2 (Oxford University Press).

TRUEMAN, E. R. (1953) "Observations on certain mechanical properties of the ligament of *Pecten*", *J. exp. Biol.* 30: 453–67.

TRUEMAN, E. R. (1966a) "Bivalve mollusks: fluid dynamics of burrowing", *Science* 152: 523–4.

TRUEMAN, E. R. (1966b) "The fluid dynamics of the bivalve molluscs *Mya* and *Margaritifera*", *J. exp. Biol.* 45: 369–82.

TRUEMAN, E. R. (1966c) "Observations on the burrowing of Arenicola marina (L.)", *J. exp. Biol.* 44: 93–118.

TRUEMAN, E. R. (1967) "The dynamics of burrowing in *Ensis* (Bivalvia)", *Proc. R. Soc.* B 166: 459–76.

TRUEMAN, E. R., BRAND, A. R. and DAVIS, P. (1966) "The dynamics of burrowing of some common littoral bivalves", *J. exp. Biol.* 44: 469–92.

TUCKER, V. A. (1966) "Oxygen consumption of a flying bird", *Science* 154: 150–1.

VOGEL, S. (1967) "Flight in *Drosophila*. III. Aerodynamic characteristics of fly wings and wing models", *J. exp. Biol.* 46: 431–43.

VRIES, H. de (1950) "The mechanics of the labyrinth otoliths", *Acta oto-lar.* 38: 262–73.

VRIES, H. de (1956) "Physical aspects of the sense organs", *Progr. Biophys. byophis. Chem.* 6: 207–64.

WALKDEN, S. L. (1925) "Experimental study of the soaring of albatrosses", *Nature, Lond.* 116: 132–4.

WARD, I. M. and PINNOCK, P. R. (1966) "The mechanical properties of solid polymers", *Brit. J. appl. Phys.* 17: 3–32.

WARNOCK, F. V. and BENHAM, P. P. (1965) *Mechanics of solids and strength of materials* (Pitman, London).

WEIS-FOGH, T. (1956a) "Biology and physics of locust flight. II. Flight performance of the desert locust (*Schistocerca gregaria*)", *Phil. Trans.* B 239: 459–510.

WEIS-FOGH, T. (1956b) "Tetanic force and shortening in locust flight muscle", *J. exp. Biol.* 33: 668–84.

WEIS-FOGH, T. (1960) "A rubber-like protein in insect cuticle", *J. exp. Biol.* 37: 889–907.

WEIS-FOGH, T. (1961a) "Thermodynamic properties of resilin, a rubber-like protein", *J. mol. Biol.* 3: 520–31.

WEIS-FOGH, T. (1961b) "Molecular interpretation of the elasticity of resilin, a rubber-like protein", *J. mol. Biol.* 3: 648–67.

WEIS-FOGH, T. (1961c) "Power in flapping flight", *in* Ramsay, J. A. and Wigglesworth, V. B. (edd.) *The cell and the organism* (Cambridge University Press).

WELCH, A., WELCH, L. and IRVING, F. G. (1955) *The soaring pilot* (Murray, London).

WIGGLESWORTH, V. B. (1965) *The principles of insect physiology* ed. 6 (Methuen, London).

WILKIE, D. R. (1956) "The mechanical properties of muscle", *Brit. med. Bull.* 12: 177–82.

WOOD, A. (1940) *Acoustics* (Blackie, London).

WORTHINGTON, C. R. (1962) "Conceptual model for the force-velocity relation of muscle (Hill's equation)", *Nature, Lond.* 193: 1,283–4.

YOUNG, J. Z. (1957) *The life of mammals* (Oxford University Press).

ZAHM, A. F., SMITH, R. H. and LOUDEN, F. A. (1928) "Drag of C-class airship hulls of various fineness ratios", *N.A.C.A. Rept.* 291: 1–16.

ZWISLOCKI, J. (1962) "Analysis of the middle-ear function. Part I: Input impedance", *J. acoust. Soc. Am.* 34: 1,514–23.

INDEX